What Color Is Newton's Apple?

Inquiry Science for Young Children

by
Jamie C. Smith, Ed.D.

Royal Fireworks Press
Unionville, New York

To my mother, Dorothea M. Smith, who helped me make mud pies and clean up spilled paints.

Dr. Jamie C. Smith has devoted her professional life to the education of young children. She is the author of *Beginning Early: Adult Responsibilities for the Education of Gifted Young Children* (published by Trillium Press). She has developed an extensive curriculum to facilitate the learning of young children, and she has trained many teachers in the education of young children.

Originally published by Trillium Press, Copyright © 1988

Royal Fireworks Press
First Avenue, PO Box 399
Unionville, NY 10988-0399
(914) 726-4444
FAX: (914) 726-3824
email: rfpress@frontiernet.net

ISBN: 0-89824-105-7

Printed in the United States of America on acid-free, recycled paper
using soy-based inks by the Royal Fireworks Printing Company
of Unionville, New York.

TABLE OF CONTENTS

INTRODUCTION

This book is a collection of activities in physics and chemistry for children from two to eight years old. The activities capitalize on the fact that children are very good at asking questions and then seeking answers through their own direct actions on objects. So the activities are designed around the idea that children learn best when they act as researchers and experimenters, working with a wide variety of materials in many different ways. The results of such experimenting are quite dynamic, and sometimes unpredictable to adults. You will quickly become a learner as well as a guide. The information which follows will help you understand how you can make the activities in this book work best for you and your children.

The Adult's Role

Learning can be enhanced by teachers and parents in several ways. First, children need an environment in which exploratory learning is both possible and valued. You need to make sure their play space is safe, can take spills and messes, and can easily be cleaned up by the children. Ideally, they should be allowed to leave things out for extended periods if necessary, even overnight. The emotional environment should be relaxed and positive, with your respect for the children and their ways of learning freely expressed. But you should also expect self-discipline from the children so that they do not endanger themselves, each other, or the materials and environment. These conditions will help you to encourage learning and creativity in the children.

Second, you can provide materials which are simple and yet which have many possible uses, such as blocks, sand, water. These materials encourage widely varied experimentation and sensory exploration. They provoke creative thinking by demanding that the children invent their own uses, responding tangibly to actions initiated by each other. To obtain the greatest benefits from open-ended materials, you need to allow plenty of opportunities to just "mess about," experiment, make noise, make a clutter, and try out ideas.

Third, you can talk to the children about their experiments. It is especially helpful to have them explain what they are doing, why, and what they expect to happen as a result. You can ask questions which encourage the children to try new actions and materials, make predictions, evaluate outcomes. These conversations about the experiments help children clarify what it is they are trying to achieve and what they may be learning through their actions. You are encouraging use of new vocabulary and articulation of ideas. You are also communicating to the children that you value their ideas and that scientific explorations are important.

Fourth, you can provide occasions when several children work together with a set of materials. Just working together, they will discuss their ideas, challenge one another's thinking and develop intellectual and social skills. Since each child brings a very personal perspective to the learning experience, conversations among children will encourage the expansion of all children's thinking. The children will encourage each other to explore and experiment more fully and to verbalize what they each experience. This is a time for you to be a careful observer, responding when the children seek your help or attention while keeping your interaction with them focused on their requests.

Fifth, you can observe your children to determine when they are ready for new or tougher intellectual challenges. If our children lose interest in a toy or material, you can help them find new ways of using it. They might use it for different purposes, combine it with different materials, or combine its parts in new ways. All these activities extend the life of the toy or material while encouraging the children to develop complex thinking strategies and creativity. Sometimes it helps if you put a toy or material away for a time, bringing it out again when the children's understandings have matured. Since you have ultimate control over the materials available to the children, it is your observations and skilled choices which will most help promote learning and creativity.

Using This Book

The activities in this book all emphasize the development of higher order thinking skills as defined by Benjamin Bloom in his *Taxonomy of Educational Objectives*.

Application involves the use of knowledge and understanding, especially in building new ideas and predicting future outcomes. Ideas applied can be more completely examined and more fully understood.

Analysis is the examination and comparison of ideas and events to find their similarities, differences and other relationships.

Synthesis takes ideas and puts them together in new ways or takes unrelated ideas and creates relationships among them.

Evaluation leads to the judgment of the worth and value of an idea or event, to a decision about "best," or other personal preference.

The activities are also designed to promoted selected thinking skills in science problem solving, creativity, and communication. Each activity addresses some skills from each area.

Science problem solving skills include:

problem finding: seeing ambiguities, incongruities and omissions in an event or situation

observation: using all five senses, movement and measurement to learn about a situation or event

hypothesis formulation: speculating on why and how a particular situation or event came to pass

hypothesis testing: investigating to decide if an hypothesis is correct

data collection: gathering specific observations about a situation or event

data organization: structuring data so that it can be understood

data interpretation: putting together specific facts to create a larger idea

solution finding: producing many solutions for the original problem

solution testing: finding out which solution works best

communication of solution: sharing the solution with others.

Creative thinking skills involve:

fluency: producing many ideas, including ''alternative trials'' or tests for hypotheses

flexibility: producing ideas of varied content or type

originality: generating unique ideas

elaboration: detailing and refining ideas.

Communication skills focus on:

verbal: expressing and interpreting ideas through oral and written language

non-verbal: expressing and interpreting ideas through physical demonstrations, illustrations, and body language

interpersonal: taking other individuals and their needs into consideration

listening: attending to the oral communications of others.

When children collaborate on an activity they also enhance their social skills and increase their moral awareness.

The activities each have nine parts: Purpose, Thinking Skills, Vocabulary, Prerequisites, Materials, Introduction, Continuation, Closure, and Learning Check.

Purpose: The purpose statement focuses on the major ideas to be presented through the activity. It is not meant to identify all possible learnings which could occur during the activity.

Thinking Skills: Each activity places particular emphasis on certain of the thinking skills defined above. These selected skills are identified (and clarified as needed) for individual activities.

Vocabulary: In each activity, certain vocabulary is likely to arise as essential or desirable. These words are listed so that you may easily introduce them into your conversations with the children. It is important that the vocabulary be introduced *after* the relevant event or situation. For example, introduce the term ''solution'' *after* the child has successfully dissolved a powder in a liquid.

Prerequisites: Many of the activities build on prior experiences and learnings which are essential to the child's successful participation. These ''prerequisites'' are included to help you recognize what a child must know to experience success. You should try to provide the prerequisite experiences before introducing the activity if you are unsure of a child's ability to succeed in the activity in question.

Materials: The lists of materials in each activity suggest the variety of items desirable for a group of children (quantities specified as necessary). In most cases the activity can be successfully conducted with fewer supplies or less variety of materials than specified. Materials suggested for later incorporation are indicated by "(optional)" (see Continuation below).

Introduction: When developing deeper thinking and researching skills, it is important for the children to be "hooked" on what they are doing. They need to be intensely interested and involved in their investigations. The Introduction provides plans for initiating the activity in such a way that the children readily become "hooked." Generally, one of three basic strategies is employed. The adult may model a possible use for the materials, or ask the child to show some possible uses. For the more advanced children, it is possible to have them plan some strategies for interacting with the materials in advance, including making predictions about possible outcomes.

The Introductions included here are geared to the nature of the materials and the developmental level of the children. They may be modified according to the children's experience and responsiveness. Too, if the children do not respond well to a verbal introduction, you can always move in as a model.

Continuation: To keep the activity flowing, it is often beneficial to interact with the children and the materials. You can assess the progress they are making and get a grasp of their experimental approach.

In order to facilitate continued explorations of the materials you might ask the children about what they are doing. It is especially helpful to use questions about application, analysis, synthesis, and evaluation related to their experiences. Exemplary questions are included with each activity plan and should be used at the appropriate time in the activity. You should generate your own questions, too, but be sure to adhere to the style of the questions, which encourages a wide range of responses and stimulates experimentation and discussion.

In order to further facilitate exploration of the materials, it may be helpful to introduce a new item which will somehow vary the experiments. These supplemental materials will be noted by "(optional)" in the materials list and there will be suggestions for their inclusion in the Continuation. You should bring out the optional materials if the children seem to be repeating themselves, losing interest, or are engaging in the activity for a second or third time.

Closure: All too often we adults find our thoughts interrupted by someone seeking our attention, and it is difficult to get our thoughts back in order. Unfortunately, we do the same thing to children by calling for clean-up or lunch right in the midst of their experiments. In order to permit the children to complete their thoughts, these activities are designed with a Closure which encourages the children to complete their thoughts. The Closures are arranged to allow completion of an experiment, discussion of experiences and/or planning for future experiments. The Closures are considered part of the activities and will help the children achieve the purposes of the activities. You need to allow ample time for the Closure, usually 5 to 10 minutes and possibly much longer, depending upon the design.

Learning Check: The Learning Check is included to help you determine if the various Thinking Skills for the activity have been achieved. Indicators of Thinking Skill behaviors are identified to help determine which have been achieved.

Selecting and Using an Activity

When moving children into exploration and experimentation as in these activities, it is wise to start gently. This is especially true for children who are used to structure and adult control. You can make the transition move smoothly by taking small steps - adding some new materials to a block area, asking children for their ideas about alternative ways to handle science text experiments, and the like. Essentially, you are asking the children to take back into their own control the materials through which they learn, so they need practice. They need gradually more control over a gradually increasing variety of materials and options for interaction.

Another basic principle for using these activities is that you should become familiar with the materials before introducing them to the children. You should be well aware of what the materials feel like, how they interact, what you can do with them. When you first use these activity plans, select activities which have materials familiar to both you and the children. As you become more comfortable with the activities you can move into new areas of learning. Just be certain that both you and the children are satisfied with your rate of progress.

The third basic principle for using these activities centers on time. The activities are designed to encourage very high level thinking. The children will need time to think about answers and possibly experiment to obtain them. They probably will not be able to and should not be expected to respond quickly to questions. If they do respond quickly with narrow answers, they need to be encouraged to dig further into the question and possibly come up with new ideas. Because these are thought-provoking strategies, you should allow extended periods for the activities, up to one hour or more for preschool and primary children and one-half hour for toddlers. (These times will vary greatly with the different activities and for different children.)

A fourth principle focuses on how children respond to questions. When you talk with young children about their ideas, they are as likely to show you as tell you what they are thinking about. Accept their demonstrated responses but also encourage them to elaborate verbally. You can help by supplying vocabulary for actions they demonstrate or details they point to. You should try as much as possible to use expansive language which encourages the children to talk and provide the ideas. Try to avoid supplying the ideas for them, even if you get very limited comments from the children themselves.

Finally, in addition to conducting a Learning Check, it is wise to evaluate the appropriateness of an activity plan for your child or group of children. You should consider such things as whether the Purpose, Thinking Skills, and Vocabulary were within reach of the children's learning abilities and whether the Prerequisites indicated were sufficiently well met to assure your children's success. You should consider if the variety of materials is stimulating yet not overwhelming for your children. You should also decide if the Introduction, Continuation, and Closure work effectively with your children.

If the activity is effective over all, it should be used with the children several times over a period of days or weeks, with modifications and intervening planning by the children, as appropriate. If it was not a success, try to determine if the activity was too hard, too easy, or too different from prior experiences. You can adjust use of it accordingly or modify it to make it more appropriate.

Where All This Leads

When we teach children using the activities in this book, we are helping them develop on several levels. As the children gain repeated experiences in making decisions, taking responsibility and interacting with materials, peers, and adults, they become intellectually and socially more independent and self-reliant as well as more sensitive to others and their environments. In-depth contact with a wide variety of materials in diverse circumstances leads to substantive understandings of the world around them. These, in turn, serve as the foundation for later academic learning.

When children develop thinking skills in science problem solving, creativity and communication, they gain control over their own learning. They become more able to initiate a learning experience, pursue it to its natural conclusion and communicate its effects. These skills build intellectual versatility, and can be used effectively with learning academic contents other than science. The basic skills of science problem solving are also employed, in some form, whenever a real-life problem emerges for the learner. Every well-made decision incorporates the collection, organization and interpretation of data, and the finding, possible testing, and communication of a solution (final decision). Larger problems will likely include the other skills as well. Use of creative thinking skills when solving a problem or making a decision helps ensure that the solution (decision) will be the best possible, accounting for the most possible contingencies and outside factors. And employing effective communication skills enriches all aspects of a person's life, from business-like exchanges in school and work settings, to family and friend relationships.

How The Book Happened

Kamii and DeVries' work triggered the idea of developing instructional strategies which would focus on the basic notions young children hold about the world. This was followed quite naturally by teacher preparation in using these instructional strategies. One result is the following collection of toddler, preschool and primary activities designed to help young children gain a richer understanding of the physical and chemical phenomena in their daily surroundings.

The individuals who designed many of the activities were Child Development majors taking their degrees at Appalachian State University in Boone, North Carolina. Their names appear below and with the specific activities they each designed.

Angela L. Blough
Johnna K. Bolick
Susan M. Brownell
Michael Lee Buie
Libbi Shaffner Dickson
Luwonna Ellis
Molly Suzanna Foster
Susan Gayle Golden
Julie Lynn Key
Laura Frazier Merritt
Julie Mullis
Susan I. Petracca
Martha A. Sweeney
Dana L. Walker
Debbie Wilson

BALLOONS

physics; toddler   designed by Susan Gayle Golden

Purpose:
To investigate the effects of gravity on balloons

Thinking skills:
Science problem solving: Observation, hypothesis formulation, hypothesis testing
Creativity:
   Originality: use of unusual actions, engaging dramatic imagination
Communication: Verbal, non-verbal, listening

Vocabulary:
Balloon, air, bounce, float, blow, light, heavy, helium

Prerequisites:
Familiarity with balloons and with ''chase the ball'' type games

Materials:
Many balloons, some filled with air, some with helium, some partly filled with water (optional), some empty; string, magic markers; large indoor area with tub or plastic wading pool
   Inflate helium balloons in advance and attach strings long enough to reach from ceiling to floor. Have other materials ready nearby.

Introduction:
Give the children the helium balloons to play with, helping them discover that they can pull on the strings to retrieve the balloons. While they play, inflate a balloon with air and slowly let it blow against the faces of interested children. Repeat this a couple of times before tying a knot in the end. Distribute this and the other air-filled balloons, encouraging comparisons with helium balloons. If there is interest, let the children draw on an empty balloon with a magic marker before you blow it up. Help the children observe the drawing as the balloon inflates.

Continuation:
Ask questions (rephrased for your children) and talk about what the children are doing and can do with the balloons. Model actions you have not observed, such as blowing on the balloons or kicking them. You may also wish to introduce the water balloons in the tub. Be sure to keep attention moving among the different types of balloons and the different actions.

NOTE: Water balloons could be saved for playing with balloons on a second occasion. If so, modify questions and vocabulary appropriately.

Application:

What will happen when we *(name of action)* the balloons?
How are you going to make your balloon move *(name direction)*?
Show *(name of other child)* how to do that!
How high *(far)* can you kick *(hit, punch, blow, throw)* your balloon?
How else can you make the balloon move?

Analysis:

How can we get the balloons down from the ceiling?
Why are these water balloons so heavy?
How come your balloon floats *(sinks, roll, bounces)*?
Why did the balloon break?
Which balloons are alike?

Synthesis:

How would you act if you were a balloon?
Move like a heavy water balloon.

Evaluation:

Which balloons are most fun to play with?
Would you like to float like a helium balloon or roll like a water balloon?

Closure:

Ask the children to pretend they are balloons, reaching high in the air and then sinking slowly to the floor. Once on the floor they can each "blow around" and find one balloon to keep. Put the child's name on it and have the child put it in a safe place.

Using a large box, ask children to collect and put in the box all the loose air-filled balloons. (You should meanwhile remove water balloons and tie remaining helium balloons somewhere safe.)

Learning check:

Introduction: Observe for exploration of materials
Continuation: Observe for explorations of and experimentation with materials; listen and observe for responses to questions
Closure: Observe for participation in dramatic play

BALLS

physics; toddler   designed by Susan M. Brownell

Purpose:
To explore the behavior of balls of several sizes and materials

Thinking skills:
Science problem solving: Observation, problem finding, hypothesis formulation, hypothesis testing, data interpretation
Creativity:
   Fluency: Production of alternative actions
   Flexibility: Variety of alternative actions
Communication: Verbal, non-verbal, listening

Vocabulary:
Round, roll, push, bounce, catch, drop, kick, hit, throw, lift (pick up), carry, big, small (little), red, yellow, blue

Prerequisites:
None

Materials:
Balls ranging in size from 3'' to 15'', several of each size, several of each primary color (total of 3-4 balls for each child); large, firmly carpeted play space; basket, box, or bag

Introduction:
Tell the children you have some balls for them and let them play freely. Observe how they interact with the balls and play with the balls yourself, keeping a careful eye for safety. Make observations about balls going around and around as they roll.

Continuation:
Talk about what the children are doing with the balls, using vocabulary and emphasizing verbs. Make certain you interact with each child. If children are interested, initiate games such as catch, find-a-ball, bounce or (kick) and chase. Ask questions, reworded for your children.

Application:
Where will the ball go when you *(name of action)* it?
How far can this ball bounce *(roll)*?
Show us how you roll on the big ball.
Find the big red *(use appropriate description)* ball.
Find a ball like this *(display)* ball.

Analysis:
How come your ball went over there?
Which is the biggest *(smallest)* ball?
Why are you *(name of action)* the ball?
Which balls are just alike?
How did you make the ball do that?

Synthesis:
How can you make yourself into a ball?
What new game can we play with the balls?

Evaluation:
Which ball would you most like to be?
What is your favorite way to move a ball?

Closure:
Take out a basket, box or bag. Have the children try to bounce, drop or throw the balls into the container. Continue commenting, encouraging and playing ''find-a-ball'' until all the balls are contained. Tell the children they will be able to play with the balls again on another day.

Learning check:
Introduction: Observe for exploration of and experimentation with materials
Continuation: Observe for continued experimentation, responses to questions; listen for
    comments about observations, responses to questions
Closure: Observe for successful participation in clean-up

BEAN BAGS

physics; toddler

Purpose:
To explore how throwing an object affects where it lands

Thinking skills:
Science problem solving: Hypothesis formulation, hypothesis testing, communication of
    solutions
Creativity:
   Fluency: Production of alternative trials
Communication: Non-verbal, listening

Vocabulary:
Bean bag, throw, here, there, catch, same, big, little, fat, skinny, farthest

Prerequisites:
None

Materials:
Lots of bean bags of various sizes, shapes and weights; large open playspace; large box

Introduction:
Take a pile of the bean bags and begin by throwing one to each child, calling the child's
name and encouraging him to try to catch the bean bag. As the children get their bean
bags, invite them to throw back to you or to throw and chase. Give them ample time to
get the feel of throwing bean bags.

Continuation:
Begin encouraging the children to try to throw in specific directions and to throw to each
other. Use vocabulary as appropriate and ask questions to stimulate thinking.

Application:
Which way will you throw your bean bag?
Where will the bean bag go?
Find the *(descriptive adjective)* bean bag and throw it.
How far will your bean bag go?
Which bean bag will you throw next?

Analysis:
Why did your bean bag go over there?
Where does it usually go?
Which bean bags are the same?
Why doesn't the bean bag go where you point?
How will you make your bean bag land over there *(indicate)*?

Synthesis:
How can you move like a bean bag?

Evaluation:
Which one can you throw farthest?
Which is your favorite bean bag?

Closure:
Take out the large box and have all the children try to throw their bean bags into the box. You will need to catch the misses and put them in the box. When all the bean bags are in the box, ask the children if they would like to throw bean bags again.

Learning check:
Introduction: Observe for participation, attention to verbal and visual cues
Continuation: Observe and listen for successful trials, attempts to demonstrate, responses to questions
Closure: Observe and listen for participation in final activity

BLENDING COLORS

physics; toddler

Purpose:
To begin creating secondary colors

Thinking skills:
Science problem solving: Problem finding, observation, hypothesis formulation, hypothesis testing, communication of solutions
Creativity:
    Fluency: Production of alternative trials
    Flexibility: Variety of alternative trials
    Originality: Uniqueness of solutions
Communication: Verbal, listening

Vocabulary:
Red (pink), yellow, blue, green, orange, purple, whipped cream, blend, mix

Prerequisites:
Prior experiences with finger painting

Materials:
Whipped cream or substitute in three separate bowls, each strongly tinted with one of the primary colors (red, yellow, blue) using food coloring; very clean, smooth table top (or clean sheet of heavy plastic); smocks, towels, bowl of warm water
    Place large dabs of each color of whipped cream by each child's place.

Introduction:
Put smocks on, then show the children to their places. Comment on the colors of the whipped cream and invite the children to finger paint with it.

Continuation:
Observe how the children work with their colors, pointing out and labeling when a child gets a color blend. If no one mixes his colors after several minutes, take some whipped cream for yourself and model making the colors blend. Also ask questions (reworded for your children) to help the children investigate things further.

Application:
What will you paint?
Which color will you use?
Find the *(name of color)*.
What will happen when you mix the *(names of colors)*?
How much *(name of color)* will you use?

Analysis:
How come the colors changed?
Do *(names of two colors)* always make *(name of appropriate secondary color)*?
How did you make a new color?
Which colors are alike?
Which color is like the one in your painting?

Synthesis:
What new painting will you make?
What new color will you blend?

Evaluation:
Which color do you like best?
Which color feels best?

Closure:
Ask the children to make one more painting. Comment on what they each do, focusing on colors, and get their comments in return. Then wash hands in bowl of water, dry and remove smocks.

Learning check:
Introduction: Observe for paying attention, exploration of materials; listen for comments
Continuation: Observe for continued exploration of and experimentation with materials, participation in color blending; listen for responses to questions, comments
Closure: Observe for participation in final activity

CLAY

physics; toddler   designed by Julie Lynn Key

Purpose:
To begin using and changing the properties of clay when shaping it

Thinking skills:
Science problem solving: Observation, solution finding, solution testing, communication of solutions
Creativity:
  Fluency: Production of alternative solutions
Communication: Non-verbal, listening

Vocabulary:
Clay, playdough, shape, mold, bigger, smaller, same, different

Prerequisites:
None

Materials:
Non-toxic water soluble clay and/or playdough, bowl of water, cookie cutters, bowls, spoons, plastic or newspaper covered tables
  Set up workspace in a comfortable location, outdoors if possible.

Introduction:
Put smocks on the children. Show the materials to them and encourage them to explore, manipulating the materials as they choose. Model patting the clay into a pancake and shaping it with fingers or cookie cutter. Observe the children to ensure they are not eating or throwing the clay. Redirect to molding if you need to.

Continuation:
Begin labeling the children's actions and asking them to identify their clay models. Ask questions and request actions (rephrased for your children) for which you can be shown and/or told answers. Be certain to talk with each child for a moment or two.

Application:
What will you do with your clay *(playdough)*?
What will happen when you *(name of action)*?

Show me how you made that.
What will happen if we add water?
What will your "cookie" be?

Analysis:
What's happening to your clay *(playdough)*?
Are these shapes the same?
Which shape is bigger *(smaller)*?
Which shapes are different?
What does the clay usually do when you *(name of action)* it?

Synthesis:
Make a big new shape to show me.
Let's make lots of little shapes.

Evaluation:
Which cookie cutter do you like best?
Please pick one shape to show our friends.

Closure:
When interest seems to wane, have each child make one large shape with his clay and playdough and put it into a bucket. Then he should wash hands and take off his smock.

Learning check:
Introduction: Observe for exploration of materials, following directions
Continuation: Observe for production of solutions which are displayed, responses to questions and requests
Closure: Observe for responses to requests

FLOAT AND SINK

physics; toddler

Purpose:
To investigate the distinguishing properties of objects which float versus objects which sink

Thinking skills:
Science problem solving: Problem finding, observation, communication of solutions
Creativity:
 Fluency: Production of alternative trials
 Flexibility: Variety of alternative trials
Communication: Verbal, non-verbal, listening

Vocabulary:
Sink, float, wood, metal, plastic, wax

Prerequisites:
None

Materials:
Dishpans of warm water; rocks, gravel, sand, bolts, nuts, washers, metal bottle tops, sponges, corks, bare wood pieces, pieces of paraffin, small candles, pieces of crayons, plastic bottle tops, plastic cubes or beads, other small plastic toys; newspapers, smocks, towels
 Set up for about 4 children at a time, arraying materials between the pans of water.

Introduction:
Put on smocks and tell the children they may play with the materials as they wish. Ask that they try to keep the water in the pans. Observe for a while, labeling when objects sink or float.

Continuation:
As the children become comfortable with the ideas of sink and float, begin asking questions to further their explorations.

Application:
What will the *(name of object)* do?
Find more wood *(metal, plastic, wax)* things.
Which are the things which float *(sink)*?
How do you make the *(name of floating object)* sink?
What floating *(sinking)* things do people use?

Analysis:
Why does the *(name of object)* float *(sink)*?
What does wood *(metal, plastic, wax)* usually do?
How are floating *(sinking)* things alike *(different)*?
How come you can't make floating *(sinking)* things sink *(float)*?
What happens to the water when something floats on it *(sinks in it)*?

Synthesis:
How would you move if you were floating on *(sinking in)* water?
How could you get something to float *(sink)* only part of the time?

Evaluation:
Would you like to be a floating thing or a sinking thing?
Which object floated best?

Closure:
Ask the children to put all of the floating objects in one location and all of the sinking objects in another. Ask them again what is alike in each pile and await responses. Then ask them to take off smocks and dry hands.

Learning check:
Introduction: Observe and listen for exploration of materials
Continuation: Observe and listen for experimentation with materials, relevant comments, responses to questions
Closure: Observe for successful participation in final activity; listen for responses to questions

MUD PLAY

physics; toddler

Purpose:
To explore how water affects the consistency of earth

Thinking skills:
Science problem solving: Problem finding, observation, solution finding, solution testing, communication of solutions
Creativity:
   Originality: Uniqueness of solutions
   Elaboration: Refinement of solutions
Communication: Verbal, interpersonal

Vocabulary:
Mud, gooey, wet, squeeze, squishy, mold, dig

Prerequisites:
Prior experiences with sand play and water play

Materials:
Ample wet sand or mud; plastic cups, small pails, jello molds, cake pans, shovels; outdoor playspace; lots of water for clean-up

Note: Be prepared for very dirty children, so a wading pool might be the best follow-up play. In any case, have the children wear minimum clothing and be prepared to have to bathe and change them completely.

Introduction:
Physically prepare children for mud play. Then take them to the play area and let them explore for themselves. Model filling a cup, packing it tight and turning over to unmold the mud. Do this several times for the various children to see. Begin labeling the children's actions and any shapes they construct.

Continuation:
Observe the children's interactions with the materials and model new actions as appropriate. Ask questions from below, rewording them as needed for your children.

Application:
What will you use the *(name of object)* for?
What are you going to make with your mud?
How big will you make your *(name of object child identified)*?
What will happen when you lift up the *(name of object)*?
How much mud will you need?

Analysis:
Which shapes are almost alike?
How come your mold did that *(indicate)*?
What happens when you *(name of action)* the mud?
What else feels like mud?
How can you get the mud to keep its shape?

Synthesis:
What new building can you make with mud?
How would you feel if you were all covered with mud?

Evaluation:
Which mold works best?
How is the mud most fun to play with?

Closure:
Ask all the children to help build one giant pile of mud molds. As each child puts on a second or third piece, thank him and aim him toward the clean-up water. Ask them each to wash out their molds and then wash themselves. Be sure you provide adult supervision for the washing. (A large tub of water to throw toys in will make clean-up easier, too.)

Learning check:
Introduction: Observe for exploration of materials, attempts to copy models, listen for comments
Continuation: Observe for experimentation with materials; observe and listen for successful parallel play, responses to questions
Closure: Observe and listen for successful participation in final activity and clean-up

SIZES AND SHAPES

physics; toddler   designed by Martha A. Sweeney

Purpose:
To investigate how objects of different sizes and shapes fit inside containers

Thinking skills:
Science problem solving: Problem finding, observation, hypothesis formulation, hypothesis testing, data collection, data interpretaion
Creativity:
   Originality: Use of unusual strategies
Communication: Verbal, listening, interpersonal

Vocabulary:
Large, small, round, change, button, squeeze, shape

Prerequisites:
Prior experience putting things into and removing things from containers

Materials:
Coffee cans (with edges smoothed), oatmeal containers, tennis ball containers, orange juice cans, medium sized boxes, paper grocery bags, laundry basket, garbage bags, tape; wooden blocks (square, medium-sized), clothespins, large buttons (too big to swallow), rolled up socks, 2½'' - 3'' diameter rubber balls, golf balls, nerf balls, sponges of various sizes, bean bags, playdough (have at least 3 of each item)
   Plan to work with 2-3 children at a time. Put a few of the materials in the boxes. Wrap up the boxes with the paper bags and then tape loosely.

Introduction:
Seat the children on the floor and let each child unwrap a box of ''goodies.'' Have a similar set for yourself. Encourage the children to examine what is in the boxes, dump out the contents and generally get involved with the materials. Make comments to encourage their involvement. As things are taken out of the box, suggest putting the materials back in. Model dropping with your own things. After a few minutes, get the children to empty their boxes and put empty boxes into the laundry basket.

Continuation:
Immediately hand each child a coffee can, adding new materials to those already being used.
Suggest actions the child can take, such as ''Drop the ball!'' Encourage the children to shake their cans and dump them out.
Add other containers as they work, putting used materials into laundry basket.
Continue to change materials, suggesting actions and asking questions (reworded for your children) as you interact with the children.

Application:
How will you make the *(name of object)* fit in you container?
What will happen when you try to put in the *(name of object)*?
What will you do if the *(name of object)* won't come out?
How will the *(name of object)* fit in if you squeeze it?
Find the things that will fit in the *(name of container)*.

Analysis:
Find the biggest *(smallest)* toy. Will it fit in the *(name of container)*?
What happens when you *(name of action)*?
Which things are nearly alike?
Which items can you fit into this container?

Synthesis:
Which of these things belong together?
Show me a new way to empty the *(name of container)*.

Evaluation:
Which is easiest *(hardest)*?
Did you have fun today?
Which toy do you want?

Closure:
Take out a large container, such as a garbage bag, for the children to put things in. This time all the children will put their materials into one container, the bag. You need to help them put things in the bag, labeling the materials and actions as you clean up. Suggest that you all put the same item in at the same time, generally making a game of clean-up. You may also want to use the original boxes and laundry basket to assist in cleaning up the materials.

Learning Check:
Introduction: Observe and listen for participation, exploration of materials
Continuation: Observe and listen for experimentation with materials and related comments, responses to questions
Closure: Observe and listen for participation in final activity

SQUIRTING WATER

physics; toddler

Purpose:
To explore how water moves when propelled with force

Thinking skills:
Science problem solving: Hypothesis formulation, hypothesis testing, solution finding, solution testing
Creativity:
  Elaboration: Refinement and detailing of actions
Communication: Non-verbal, interpersonal

Vocabulary:
Squirt, swirl, drops, wet, fill, colored, red, blue, yellow

Prerequisites:
None

Materials:
Dish pans of warm water, large bowls of water intensely colored with red, yellow, and blue food coloring; turkey basters, eyedroppers, clean medical syringes of several sizes, plastic bottles with squirt tops; smocks, newspapers, towels

Set out materials in an area where food color stains will not be a problem. Be aware that the children's fingers (and yours) could become tinted temporarily by the intensely colored water.

Introduction:
Put smocks on and ask the children if they would like to try squirting the water in the pan. Show them the squirters (but keep colored water aside) and observe as they try to operate the various devices. You may need to model how something works and you may need to remind children to squirt the water <u>into</u> the pans.

Continuation:
As the children become proficient in operating the various squirters, introduce the colored water and tell them they may squirt it into the large pans. Ask questions (reworded for your children) and encourage the children to try new actions.

Application:
What will you do with this squirter?
What will happen when you squirt *(name of color)* water?
What will happen if you squirt this under the water in the pan?
Show me how you *(name of action)*.

What will you try next?

Analysis:
How come the water looks like that *(indicate)*?
How did you make the color change?
Why is the water swirling?
Which squirters are most alike *(different)*?
What usually happens when you *(name of action)*?

Synthesis:
Make a design with your colored water.

Evaluation:
Which squirter do you like best?
Do you like to squirt the water?

Closure:
Ask the children to squirt all of the colored water into the large tub. Then have them each take off smocks and dry hands.

Learning check:
Introduction: Observe for exploration of and experimentation with materials
Continuation: Observe for continued experimentation, successful parallel play, responses to questions and suggestions
Closure: Observe for cooperation, successful clean-up

STACK AND BALANCE

physics; toddler

Purpose:
To construct stacks of objects which retain their balance

Thinking skills:
Science problem solving: Hypothesis formulation, hypothesis testing, data interpretation
Creativity:
   Fluency: Production of alternative trials
   Flexibility: Variety of alternative trials
Communication: Verbal, non-verbal, interpersonal

Vocabulary:
Stack, balance, fall, tall

Prerequisites:
Prior experiences playing with blocks

Materials:
Empty cardboard boxes with lids taped shut, hollow cardboard blocks, hollow plastic blocks, foam blocks, wooden unit blocks, small wooden blocks; large playspace with firm floor covering

Introduction:
Bring the children to the play area and show them the large hollow blocks and boxes (keep wooden blocks aside). Ask them to make towers as tall as they can make them. Limit your interaction during their initial explorations.

Continuation:
Observe to see if children are stacking to achieve tall structures. If they are not, model stacking with several of the materials. If the children are stacking well, introduce the wooden blocks for them to try. Comment on their work, using the vocabulary and ask questions to further their investigations.

Application:
Where will you put this block?
How tall will you make your tower?
What will happen when you add one more block?
What will you build next?
When will your tower fall?

Analysis:
Why did your tower fall?
How did you make the blocks balance?
When do blocks balance?
What makes towers stay standing?
How are these two towers different *(the same)*?

Synthesis:
How can you put all these blocks together into one giant tower?
Design the best tower you can.

Evaluation:
Which blocks balance the best *(worst)*?
Whose tower made the loudest noise when it fell?

Closure:
Involve the children in building one giant construction which they all can then knock over. Once it is knocked down, have them pile their blocks into boxes until all the blocks are off the floor.

Learning check:
Introduction: Observe and listen for exploration of and experimentation with materials and related comments, parallel play
Continuation: Observe and listen for continued experimentation, responses to questions, parallel and possibly some cooperative play
Closure: Observe and listen for participation in final activity

TEXTURES

physics; toddler   designed by Susan I. Petracca & Michael Lee Buie

Purpose:
To explore the textures of a variety of common substances

Thinking skills:
Science problem solving: Observation, communication of solutions
Creativity:
   Originality: Unusual uses of materials
Communication: Verbal, non-verbal

Vocabulary:
Sticky, smooth, slimy, crunchy

Prerequisites:
None

Materials:
Sticky - marshmallow topping, butterscotch topping, smooth peanut butter
Slimy - cooked okra, semi-solid jello, over-cooked and chilled spaghetti
Smooth - cool whip, whipped margarine, smooth yogurt
Crunchy - dry textured noodles, potato chips, dry coarse cereal
Waxed paper, smocks, tape, sponges; bowls of warm water for clean-up; smocks
   Select two of the edible materials from each group, using one from each group to begin
with. Cover table with wax paper placemats.
   This activity should be restricted to four childern at a time.

Introduction:
Put smocks on the children. Using the four substances, give each child one and suggest
he make designs or ''paint'' with it on the waxed paper. Leave the containers on the
table. Begin labeling substances and textures after the children have explored for a few
minutes, making other comments as appropriate.

Continuation:
Tell children they may use the other materials on the table and begin introducing the
remaining four substances.
   Suggest that they may use these to ''paint'' with, too. Make comments and ask questions
(rephrased for your children) to stimulate the children's involvement.

Application:
What will the *(name of substance)* feel like?
What will you do with the *(name of substance)*?
Show me how you mixed the *(names of two substances)*.
Show us how to make a design with the *(name of substance)*.
What other parts of your arms *(bodies)* can you use to make a design?

Analysis:
How does the *(name of substance)* feel?
How do the *(names of two substances)* feel when they are mixed together?
What happened when you mixed the *(names of substances)*?
How are the feelings alike *(different)*?
How come this feels *(appropriate adjective)*?

Synthesis:
What new texture *(color)* will you make?
What will you paint?

Evaluation:
Which feels *(tastes, looks)* the best?
Which other things would you like to try?

Closure:
Remove the bowls of substances and put down the bowls of warm water. Have each child wash his hands in the bowls while you use a face cloth to clean elbows, faces, etc. Make sure the children dry off well. Talk more about textures, especially the texture of the water, while you clean up.

Alternative:
Select substances of more similar textures, such as whipped cream, yogurt, and jello, choosing different colors and flavors. Encourage finer discriminations among textures, modifying vocabulary and questions appropriately. You can also begin to incorporate color blending.

Learning check:
Introduction: Observe for exploration of materials; listen for comments
Continuation: Observe for exploration of materials, "paintings," and other displays; listen for comments, responses to questions
Closure: Listen for generalizations

WATER PLAY

physics; toddler

Purpose:
To explore how water reacts to different actions

Thinking skills:
Science problem solving: Observation, communication of solutions
Creativity:
   Fluency: Production of many actions
   Originality: Uniqueness of actions
Communication: Non-verbal, interpersonal

Vocabulary:
Pour, splash, bubble, dribble, big, little, into, onto, in, waterfall

Prerequisites:
None

Materials:
Tubs of warm water (can use wading pool); plastic cups and bottles of a variety of sizes; smocks if desired, towels
   Place materials out in a waterproof area. Plan on up to 6 children at one time.

Introduction:
Put smocks on or undress children to underpants (clean diapers), whichever is preferred. Bring them to the water play area and model pouring water from a cup. Encourage the children to try, reminding them to keep the water in the tubs.

Continuation:
As the children get involved, begin labeling their actions. Model pouring from one container into another, splashing and other actions the children can imitate. Also ask questions to provoke further explorations.

Application:
What will happen when you *(name of action)* the water?
What are you going to do next?
What will happen when you pour the big bottle into the little cup?
Show us how you made the big *(little)* splash.
Show us how to make a big waterfall *(little dribble)*.

Analysis:
How come you made a big splash?
What happens when you pour out a bottle *(cup)*?
Where does all the water go?
How do you make a big *(little)* splash?
How come the cup overflowed?

Synthesis:
How can you make one big splash all together?
What other toys should we bring to the water play next time?

Evaluation:
Which way makes the best splashes?
Which bottle makes the biggest waterfall?
Did you like splashing *(making waterfalls)*?

Closure:
Tell the children that now they can make a big splash all together. Tell them to hit the water really hard when you say "three," then count 1, 2, 3. Then begin drying off the children with those dry going to a new activity, those wet staying with the water play until dried off. (If some children get upset by the splash, dry them off first.)

Learning check:
Introduction: Observe for exploration
Continuation: Observe and listen for exploration of and experimentation with materials, parallel play, responses to questions
Closure: Observe for participation in final activity, working as part of group

BALANCING

physics; n-k   designed by Laura Frazier Merritt

Purpose:
To maintain one's balance when the body posture is changed and when additional weights are used

Thinking skills:
Science problem solving: Hypothesis formulation, hypothesis testing, solution finding, solution testing, communication of solutions
Creativity:
  Fluency: Generating alternative solutions
  Flexibility: Producing a variety of solutions
  Originality: Finding unique solutions, engaging dramatic imagination
Communication: Verbal, non-verbal, interpersonal

Vocabulary:
Balance, weight, posture

Prerequisites:
Repeated experiences working on a balance beam without weights; experiences balancing blocks and other objects

Materials:
Balance beams (one for each supervising adult to work at), 20-30 feet of washline, floor mats; bean bags, large and small tin cans and plastic containers, hula hoops
  Arrange the balance beams and rope in a large play area; have other materials near each beam and the rope.

Introduction:
Ask the children if they remember how it felt to walk on the balance beams. Discuss their reactions. Then tell them that today they will try something new, carrying objects across the balance beam. (You may want to set up a dramatic scene in which they are carrying needed supplies across a bridge high over the water. Be careful not to make it sound too scary or some children will be hesitant to try.)
  Ask the children how they think carrying an object will feel on the balance beam. Tell them if it is too scary they can try walking the rope first. Also, tell them they can practice balancing their objects while they await their turns, maybe standing on one foot or tip toe. Now get the group started, with you (and your aide) carefully spotting at the balance beams.

Continuation:
As the children become assured of their abilities to balance holding an object, you (and your aide) should ask questions to provoke their thinking about the balancing process. You should be careful to use the vocabulary as you work with the children.

Application:

How do you use different parts of your body to help you balance?
How can you use different parts of your body to help you carry things?
What will happen to your balance if you *(name of action)*?
What must you do to get the hula hoop to balance?
Please show us the best way to carry a *(name of object)*.

Analysis:

How will the *(name of object)* make balancing easier *(harder)*?
Where should you place the *(name of object)* to make balancing easier?
How is balancing with an object different from balancing with empty hands?
What happens when you try to balance on the balance beam *(rope)*?
How is balancing the hula hoop like *(different from)* balancing your body?

Synthesis:

What could you do to your body to make it easier to balance while carrying things?
Which combinations of objects make carrying easier *(harder)*?

Evaluation:

What was the easiest *(hardest)* way to carry objects across the balance beam *(rope)*?
What was the most unusual way you thought of to carry things?

Closure:

When the children have a good idea of how balance works, ask each child to select one object. Then, without using their hands, they should each carry their objects to a specified location and place them in a pile. As the objects are placed in a pile, ask the children to sit in a circle and share how they balanced things today. Encourage them to share their observations and reinforce the vocabulary.

Learning check:

Introduction: Observe and listen for possible explanations, attempts to test explanations
Continuation: Observe and listen for tests of explanations, generation and testing of multiple solutions, beginning of sharing solutions verbally and non-verbally, assisting one another
Closure: Observe and listen for explanations, demonstrations of success, assistance to peers

BEAN BAGS

physics; n-k

Purpose:
To investigate how trajectory is affected by weight, shape and how the object is propelled

Thinking skills:
Science problem solving: Observation, hypothesis testing, data collection, data interpretation
Creativity:
  Fluency: Production of alternative trials
  Originality: Uniqueness of alternative trials
Communication: Verbal, non-verbal

Vocabulary:
Light, heavy, symmetrical, irregular

Prerequisites:
Prior experiences aiming at targets

Materials:
Bean bags of a variety of weights and shapes; board with holes large enough to fit bean bags through easily, empty buckets, empty boxes; open playspace

Introduction:
Set up materials for parallel play. Bring a group of children together and show them the bean bags. Ask if they have ever played with bean bags before and how they used the bean bags. Tell them that this time they will be able to throw the bean bags at targets. Ask them to name some different ways to throw the bean bags, then tell them each to find a workspace and begin experimenting with their ideas.

Continuation:
Circulate among the children, observing their strategies and asking them pertinent questions. Help them focus on the differences among the bean bags and how accurately they are able to aim the different bean bags.

Application:
What will happen when you *(name of action)* this bean bag?
How will this *(indicate oddly shaped or heavy bean bag)* one move when you throw it?
How else can you get the bean bag to hit the target?
What are some other games we use our aim in?
What would happen if the bean bag were much bigger *(smaller)*?

Analysis:

Why did your bean bag land over there *(indicate)*?

What usually happens when you *(name of action)* the bean bag?

What usually happens when you throw the *(indicate oddly shaped or heavy bean bag)*?

Why is it harder *(easier)* to aim the oddly shaped bean bags?

How do you decide where to stand before you *(name of action)* the bean bag?

Synthesis:

What kind of new game can you invent using bean bags?

How would you design the best bean bag in the world?

How can you improve the target you are using?

Evaluation:

Which is the easiest *(hardest)* bean bag to throw *(or other action)*?

What is the best way to get a bean bag to hit the target?

Closure:

Ask the children to throw all of their bean bags into their targets and come sit in a circle. Discuss what they tried and which bean bags were easier to aim. Ask them how they would want to play with the bean bags next time you have them out. Keep a record of their suggestions.

Learning check:

Introduction: Listen for relevant comments; observe for exploration of materials

Continuation: Observe and listen for experimentation with materials, responses to questions

Closure: Observe for participation; listen for comments and suggestions

NOTE: To adapt this activity for ring toss, use a variety of ring shapes, sizes and weights with appropriate targets and modify the questions to reflect the change in materials.

BLOWING

physics; n-k   designed by Angela L. Blough

Purpose:
To determine what characteristics make objects easier or harder to move by blowing on them

Thinking skills:
Science problem solving: Hypothesis formulation, hypothesis testing, data collection, data organization, data interpretation
Creativity:
   Originality: Finding unusual solutions
   Elaboration: Refinement of strategies
Communication: Verbal, non-verbal, interpersonal, listening

Vocabulary:
Light, heavy, soft, stiff, weight, friction, suction, "blowable"

Prerequisites:
Experience blowing bubbles through a straw

Materials:
Pieces of light paper (paper punch-outs work well), construction paper, pencils, straws, cotton balls, plastic bottle tops, facial tissues, metal jar lids, wood chips, spools, blocks, nails, yarn, dry leaves, marbles, ping pong balls, rubber balls, small wheel toys; large pieces of fine and coarse sandpaper; several straws for each child; smooth open floor space or large table tops; experience chart materials

Introduction:
Lay out the materials on a table top and ask the children to look at them carefully. Ask them which items they think would be easy to blow across the floor or table top and why. As they name items, make a list of "easy" and another of "hard" on the chart. If the children omit any items, ask about those items specifically.

   Break the children into groups of 2-3 to test out their predictions. Remind them to share the materials among the groups.

Continuation:
Use the questions below and others to help the children further their thinking. If they are very absorbed in the blowing, wait until they are willing to tend to your comments before intervening. You may have to reserve the questions for the Closure since the children will have straws in their mouths much of the time. You should introduce the sandpaper as a surface for blowing things on.

Application:
What will happen if you blow with more than one straw at a time?
How can you move the *(name of object)* in a different way?
What will happen if you move the straw closer *(farther away)*?
What would happen if the surface of the floor *(table)* were rough?
How hard will *(did)* you have to blow to make the *(name of object)* move?

Analysis:
How come the *(name of object)* moved like that?
How are the easy- *(hard-)* to-blow objects alike *(different)*?
What usually happens when you try to blow paper *(wooden, metal, plastic)* things?
What usually happens when you blow with several straws?
How does the size *(weight, shape)* of *(name of object)* affect the way it can be blown?
What affect does a rough surface have on how "blowable" things are?

Synthesis:
How could you use all the children in our class to make a giant blowing machine?
What would you name that machine?

Evaluation:
Which objects were easiest *(hardest)* to blow?
If you were a *(name of object)*, would you enjoy *(not)* being "blowable?"

Closure:
When it is nearly time to complete the activity, place an empty box on its side on the floor (or on the floor by the end of the table). Ask the children to work together to blow everything into the box. If they can't blow something, ask that they try another way to move it without using their hands or teeth (this may get them thinking about suction for a future activity). Once all the materials (except the used straws, which need to be discarded) are in the box, bring the group together by the experience chart. Ask the children to identify which predictions were correct and which were not. Discuss "why" when there is disagreement about how well an object blew.

Learning check:
Introduction: Observe and listen for predictions and their rationales
Continuation: Observe for tests of predictions, refinement of strategies, sharing; listen for comments regarding tests, responses to questions and directions
Closure: Observe for alternative strategies; listen for comparison of findings with predictionsi

BLOWING BUBBLES

physics; n-k

Purpose:
To control the production of bubbles in various media

Thinking skills:
Science problem solving: Problem finding, observation, hypothesis formulation, hypothesis testing
Creativity:
  Originality: Unusual uses of materials
Communication: Verbal, listening, interpersonal

Vocabulary:
Bubble, blow, suds, foamy, exhale, inhale, burst

Prerequisites:
Prior water play experiences using squirt bottles

Materials:
Dish pans of warm water with lots of dish detergent (or half with, half without detergent and extra detergent available); straws, turkey basters, squirt bottles, eyedroppers; bubble liquid, bubble pipes, bubble rings; towels, smocks; waterproof workspace

Introduction:
Gather the children together and ask if they have ever blown bubbles before. Talk about what they did for a few minutes. Then tell them you have a bunch of things they can use for blowing bubbles, so they should put on smocks. Let them experiment freely for a while.

Continuation:
As you observe, note the strategies the children use, pointing out the more successful or unusual techniques. If they are missing some obvious uses of the materials, model for the children. Use vocabulary and ask questions as appropriate.

Application:
How will you use the *(name of object)*?
What will happen if you squeeze the turkey baster *(eyedropper)* in *(over)* the water?
How can you make bubbles with the bottles?
How many bubbles will you make?
What will happen if you blow on the bubbles?

Analysis:

How does the *(name of object)* help you make bubbles?
Why are some of the bubbles large and some small?
Why can *(can't)* you make bubbles in this *(indicate)* pan?
What is inside of a bubble?
How does a bubble ring work?

Synthesis:

What equipment would you need to get one giant bubble?
How would you feel *(move)* if you were a bubble?

Evaluation:

What made the best *(worst, biggest, most, smallest)* bubbles?
Would you rather be a small bubble or a big bubble?

Closure:

After a two minute warning, bring all the children together to pop the bubbles you blow with the bubble ring. Spend about 3-5 minutes popping bubbles before you sit down to talk for a few more minutes. (You might give each child a bubble to try to take with him to the group discussion.) During the discussion, talk about where the bubbles come from and what things are needed to make bubbles. If the children raise the issue, discuss the differences between soap bubbles and soda bubbles.

Learning check:

Introduction: Listen for participation in discussion; observe and listen for exploration of materials

Continuation: Observe and listen for experimentation with materials, responses to questions, sharing materials

Closure: Observe for taking turns; listen for contributions to discussion

BOARDS AND ROLLERS

physics; n-k

Purpose:
To user rollers as primitive wheels for moving heavy objects over distances

Thinking skills:
Science problem solving: Problem finding, observation, hypothesis formulation, hypothesis testing
Creativity:
  Fluency: Production of alternative trials
  Flexibility: Variety of alternative trials
  Elaboration: Detailing of alternative trials
Communication: Verbal, interpersonal

Vocabulary:
Roller, balance, diameter, smooth, rough, bumpy

Prerequisites:
Prior experiences with wheel toys

Materials:
Boards of various lengths and widths, thick enough to hold a child's weight; variety of rollers including thick wooden rods, metal pipe, plastic pipe, with numerous pieces of each diameter; empty boxes large enough for a child or two to sit in; 5-10 ft. long pieces of rope; large indoor playspace

Introduction:
Work with 4-5 children at a time. Show the children the materials and ask them how they think they can use the materials to make a train for themselves. You may want to remind them that these things are for building with. Take suggestions from the children, encouraging detailing of their ideas. Then allow them to experiment to see what will actually work.

Continuation:
As the children work on building a successful train, ask questions and introduce vocabulary as appropriate.

Application:
How can you use these things to make a train?
What else can you use them for?
What will happen if you mix up all the different sizes of rollers?
What will happen if you put a board on only one roller?
What are some other combinations of boards and rollers?

Analysis:

How come the rollers *(boards)* acted that way *(indicate situation)*?

What happens when you mix all the sizes of rollers together?

Why do you have to arrange the rollers and boards in certain ways to make a train?

What makes one train ride different from another?

Which boards *(rollers)* usually work better *(less well)*?

Synthesis:

How would you design a train if all the rollers were different sizes?

What other kinds of materials could you use to improve your train?

Evaluation:

Where would you like your train ride to take you?

If you could pick only one way to use boards and rollers, which would it be?

Which train gave you the best ride?

Closure:

Ask the children to make one long train ride, to be their last. Then help them take the ride all together. Once their ride is over, assign them to picking up boards, rollers and ropes and putting them in the boxes. You should then spend several minutes planning for a future session with boards and rollers, asking what different materials the children would like to add.

Learning check:

Introduction: Observe and listen for exploration of materials, cooperation; listen for suggestions and relevant comments

Continuation: Listen for responses to questions; observe and listen for experimentation with materials, cooperation

Closure: Observe and listen for collaboration, participation in clean-up, participation in discussion

BOUNCING BALLS

physics; n-k   designed by Johnna K. Bolick

Purpose:
To investigate the trajectories of different types and sizes of balls

Thinking skills:
Science problem solving: Problem finding, solution finding, solution testing
Creativity:
  Fluency: Production of numerous responses
  Elaboration: Refinement of responses
Communication: Non-verbal, verbal, listening

Vocabulary:
Positional prepositions, adverbs, and adjectives, such as on, into, high, low; bounce, roll

Prerequisites:
Basic experiences aiming balls at targets

Materials:
Large and small rubber balls, tennis balls, basketballs, racket balls, ping pong balls, tiny rubber balls, superballs, nerf balls, soccer balls; large cardboard boxes; large safe indoor or fenced outdoor playspace

Introduction:
Put the balls out where they are accessible to the children. Allow the children to begin by playing as they wish, drawing attention to children who are bouncing balls or playing catch by bouncing them to one another. Ask questions like, ''Which ball can you bounce the highest?'' to get the children to focus on bouncing.

Continuation:
Introduce the boxes. Ask questions to provoke using the boxes as targets and to encourage thinking about how the various balls move through space, especially when being bounced.

Application:
How can you use this box with the balls?
What will happen if you try to hit the box from far away *(near by)*?
How can you keep the ball from bouncing back out of the box?
How many bounces will *(did)* it take to get the ball into the box?
Where should you stand so it is easy *(hard)* to get the ball into the box?

Analysis:

How come your ball bounced like that?

How should you throw the ball differently to get it into the box?

Why do some balls bounce back out of the box?

Why is it hard *(easy)* to bounce the ball into box from where you stand?

How will it be different if you sit *(lie)* down to bounce the ball?

Synthesis:

What game could we play using the boxes and balls?

How could we put all the balls and boxes together into one new toy?

Evaluation:

Which balls bounced the best *(worst)*?

Was it easy or hard to get the balls into the boxes?

What is the trickiest way you can make a ball bounce into the box?

Closure:

After the children have played for an ample amount of time, announce a short contest. The rule is that once a ball is in a box, it must stay there. Have the children see how few bounces they need to get balls into boxes. Make sure every child gets declared a winner. Once all the balls are bounced into boxes, ask the children to sit in a circle and share a little about what they did. Ask them to make generalizations about the large and small balls, hard and soft balls, etc. Also ask if they would like to play with bouncing balls again.

Learning check:

Introduction: Observe for attempts to bounce balls and predict trajectories

Continuation: Observe for experimentation with materials, solution production, increasing accuracy, demonstrations of ideas

Closure: Observe for demonstrations of successful solutions, following instructions; listen for comments related to solutions

COLOR MIXING

physics and chemistry; n-k/primary   designed by Michael Lee Buie

Purpose:
To create shades of colors in a variety of media

Thinking skills:
Science problem solving: Problem finding, observation, data collection, data organization, solution finding, solution testing, communication of solutions
Creativity:
   Fluency: Production of alternative trials
   Flexibility: Variety of alternative trials
   Originality: Uniqueness of alternative trials
   Elaboration: Refinement of alternative trials
Communication: Verbal, non-verbal

Vocabulary:
Shade, tint, pale, dark, intense, intensity, media

Prerequisites:
Knowledge of primary and secondary colors and how they are related

Materials:
Finger paints, water colors, powdered paints, all of various colors including black and brown; food coloring, crayons, colored and white chalk, colored playdough; cool whip, white glue; baby food jars filled with water, eyedroppers, paint brushes, colored and white paper, white egg cartons, newspapers
   Set up materials for groups of 4-5 children.

Introduction:
Working with the whole group of children, briefly review the primary and secondary colors. Divide the children into groups of 4-5 and ask each group to get newspaper to cover its work area. Then, either distribute or ask each group to pick up its set of materials. Ask the children how they can change the colors of the paints by using the materials they have at their work areas. If they do not answer or begin mixing colors, model by making a color darker or by blending two colors. Have the children keep records of what they try, if they write fluently.

Continuation:
As the children work, circulate among the groups and ask questions about what they are doing and finding. Discuss with them what they would like to accomplish and encourage them to make predictions about the combinations of materials. Use the vocabulary as relevant.

Application:
How could you use these materials to make *(name of blended color)*?
What will you have to do to lighten *(darken, intensify)* a color?
How much *(name of substance)* will you need to use to make a change?
What will you have to do to make the *(name of colored substance)* match this crayon *(other paint)*?
How will you go about changing the colors of the crayons?

Analysis:
What happened when you mixed *(name of two or more substances/colors)*?
What qualities do the colors *(media)* share?
How are the colors *(media)* different?
What usually happens when you blend *(names of colors)*?
What did you generally have to do to lighten *(darken)* your colors?

Synthesis:
What does this new color make you think of?
How can you display the various colors you are blending so that you show how they were created?

Evaluation:
Which combinations of media worked best *(least well)* for lightening *(darkening, intensifying)* the colors?
Which new color is your favorite?
Did the new colors turn out as you expected?

Closure:
After the children have had some success making colors, open the water play area (if you have one) to use during clean-up (or use buckets of water or the in-class sink).
Ask each group to clean up its work area, setting aside to dry any paintings they wish to keep. When clean-up is complete, bring the children together to discuss what they tried and what they learned. Make an experience chart as appropriate.

Learning check:
Introduction: Listen for comments and inquiries; observe for exploration of materials, record keeping
Continuation: Observe and listen for suggestions, experimention with materials, communication about successes and failures; observe for continued record keeping
Closure: Listen for pertinent comments; observe for displays of paintings

FRICTION

physics; n-k

Purpose:
To investigate the effect friction has on moving objects

Thinking skills:
Science problem solving: Problem finding, observation, hypothesis formulation, hypothesis testing
Creativity:
    Fluency: Production of alternative trials
    Originality: Uniqueness of trials
Communication: Verbal, interpersonal

Vocabulary:
Friction, resistance, difficult (hard), easy

Prerequisites:
Prior experience playing with boxes and with wagons

Materials:
Empty cardboard boxes, some large enough for a child to sit in; wheel toys, some large enough for children to ride in or on; old wooden and plastic blocks, tennis balls; several distinct grades of sand paper (ranging from very fine to very coarse); smooth wide boards, blocks for making ramps, lengths of rope; smooth and rough floor surfaces (grass and concrete or asphalt work very well)

Attach the sand paper to the wide boards so that there are 4-5 inches of clear board between each grade of sand paper. The sand paper need not run the entire length of the boards. Put out other materials, including loose pieces of sand paper, in a convenient work area.

Introduction:
Bring the children together and show them the materials. Ask them how it feels to push or pull heavy loads in boxes or wagons. Encourage them to discuss any experiences they remember. Ask them if some ways of moving things are easier than others. Await responses, then tell them they are free to find out whatever they can about moving things.

Continuation:
Help the children get fully involved in investigating friction by modeling some of the ways things can be moved using the materials at hand. Also begin asking questions and introducing vocabulary.

Application:

How will you use these other materials?

What will happen when you try to slide the block on the smooth board *(rough sand paper, grass, concrete, etc.)*?

What are some ways that people use friction?

What will happen when you push *(pull)* the empty *(full)* box?

How will you keep the block *(box)* from sliding too fast?

Analysis:

Why is this *(indicate)* surface easier *(harder)* to slide things on?

What are some similarities *(differences)* in the surfaces you tried?

What usually happens when you pull *(push)* objects on a rough *(smooth)* surface?

How are pushing and pulling different *(similar)*?

What causes friction to occur?

Synthesis:

How would you design a vehicle *(road surface)* so that it makes only a little friction?

How can you change friction so that it really helps people a lot?

Evaluation:

Which combination seems to produce the least *(most)* friction?

What is the most important way friction helps *(hurts)* people?

If you had to slide a heavy box up a hill, would it be better to have too much friction or too little friction?

Closure:

Ask everyone to push or pull the materials into one location. When they all come together, ask them to try to judge whether they had a lot or only a little friction to work against. Also ask them to compare their last experience with what they remember from their experimenting. Spend a few minutes sharing these thoughts and reinforcing vocabulary.

Learning check:

Introduction: Listen for contributions to discussion; observe for exploration of materials

Continuation: Listen and observe for experimentation with materials, cooperation, responses to questions

Closure: Observe and listen for cooperation in final activity, participation in discussion

GRAVITY

physics; n-k

Purpose:
To investigate the effects of gravity on a variety of objects

Thinking skills:
Science problem solving: Problem finding, observation, data collection, data interpretation, communication of solutions
Creativity:
   Fluency: Production of alternative trials
   Flexibility: Variety of alternative trials
   Originality: Unusualness of alternative trials
Communication: Non-verbal, listening

Vocabulary:
Gravity, drop, float, drift, aim, weight, surface, shape

Prerequisites:
None

Materials:
Clothespins, empty soda cans (some crushed), feathers, sheets of paper, 1 in. wooden cubes, small blocks, small balls, cotton balls, crayons or short pencils; 1 gal. plastic mayonnaise or pickle jars (wide-mouthed) and/or 2-3 lb. coffee cans with edges smoothed, one gal. milk jugs without lids (optional); chairs or climbing equipment to stand on; experience chart materials

Introduction:
Bring together the children and model dropping several of the heavier objects into a wide-mouthed container. Ask the children if they would like to try, too. Give them turns and start setting out additional materials for parallel play.

Continuation:
Observe their progress in improving aim and in trying out all the different materials. When they are ready, introduce the possibility of standing on something tall to see if height from the target makes any difference. Show them specifically which things they may stand on when trying the greater heights. Also begin introducing relevant vocabulary and asking questions to further their thinking.

Application:
What will happen when you try to aim the *(name of object)*?
How do you plan to play with your materials?
Do you expect it to be easier, harder, or just the same from up high?
What will happen when you wad up the paper?
How will you remember what you have already tried?

Analysis:

Why did the *(name of object)* land over there?

How are dropping things from down low and up high alike *(different)*?

What usually happens when you try to drop a *(name of object)* into a *(name of container)*?

Why is it harder to get objects into the milk jug?

Why do these *(indicate)* objects behave in similar ways?

Synthesis:

What kind of game could you develop using dropping objects?

How would you design the perfect object *(container)* for your dropping game?

Evaluation:

Which objects are easiest *(hardest)* to get into the containers?

Which materials would you choose for a dropping game?

Closure:

Ask the children to drop objects into their containers until all the objects have been tried one last time. They should then bring the containers and sit in a circle. Ask them to dump out whatever is in the containers so you can make a list (can be pictures) for each child of what he succeeded in getting into his container. Ask the children which of the objects were in the most containers. Also ask the children what other things they would like to try dropping, making a list of suggestions.

Learning check:

Introduction: Observe for attention to modeling and follow-up instructions, exploration of materials

Continuation: Observe for experimentation with materials, continued following of instructions; listen and observe for responses to questions

Closure: Observe for following instructions, participation in final activity

INCLINE PLANES

physics; n-k   designed by Susan M. Brownell

Purpose:
To construct incline planes to help things move up and down easily

Thinking skills:
Science problem solving: Hypothesis formulation, hypothesis testing, communication of
    solutions
Creativity:
   Fluency: Production of alternative trials
   Flexibility: Variety of alternative trials
Communication: Verbal, non-verbal, listening

Vocabulary:
Incline, ramp, slant, slope, steep, shallow, smoothly, slide (verb), incline plane

Prerequisites:
Experience building with blocks and playing with wheel toys

Materials:
Boards (1 ft. x 2-3 ft., 2 ft. x 2-3 ft., 2 ft. x 2 ft. - one of each for each child); numerous
building blocks, set(s) of wooden stairs; several large wooden trucks or cars, 2-3 small
cars per child, large cardboard box (for child to sit in), 3'-4' rope; experience chart
supplies; large playspace
   This activity should be set up for 3-4 children at a time.

Introduction:
Bring the children to the materials (except box and rope) in the playspace and ask them
what they can do with these things. Allow them to play freely for a while. Then ask
them how they could drive a large truck/car to the top of the stairs. If they bump the
truck/car on the steps, ask what they could do to help it roll smoothly.

Continuation:
As the children experiment with ways to solve this problem and are beginning to work
successfully with ramps, use questions to provoke further experiments. You should also
introduce the box and rope, asking how the children might want to use the new materials.
Employ vocabulary words as appropriate.

Application:
What are some different ways you can use these materials?
How can this board help you get a truck *(car)* up and down the stairs?
What would happen if you tried to slide the *(name of object)* down *(up)* the stairs?
What will happen to the *(name of object)* on a steep *(shallow)* ramp?
What can you do with this rope and box?

Analysis:
How are these inclines *(ramps)* alike *(different)*?
What usually happens on the steeper *(shallower)* ramps?
Why did your car go farther *(faster, slower)*?
Why do we need incline planes?
How are stairs and an incline plane alike *(different)*?

Synthesis:
How would you design a race track using incline panes?
What would it feel like to BE an incline plane?

Evaluation:
What was the easiest thing to move on the incline plane *(ramp)*?
If you could choose one toy to ride on the ramp, what would it be?
What is the most important use for incline planes?

Closure:
After the children have worked with a variety of ramps tell them that you will conduct a race for clean-up. Choose (or set up) a wide ramp. Have the children each place one small car at the top of the ramp and race the cars into the box. When all the cars are done racing, ask the children to pile up all the other materials while they think about what to do with the ramps next time. Take a few minutes to hear their suggestions, writing them down on an experience chart.

Learning check:
Introduction: Observe and listen for responses to questions and suggestions
Continuation: Observe for experimentation with materials, demonstration of understandings; observe and listen for responses to questions
Closure: Observe for responses to suggestions and requests

INNERTUBES

physics; n-k   designed by Libbi Shaffner Dickson

Purpose:
To use the shape and elasticity of innertubes to obtain desired actions

Thinking skills:
Science problem solving: Problem finding, observation, solution finding, solution testing, communication of solutions
Creativity:
   Fluency: Production of alternative trials
   Flexibility: Variety of alternative trials
   Originality: Unusualness of alternative trails
Communication: Verbal, non-verbal, interpersonal

Vocabulary:
Elastic, springy, balance

Prerequisites:
None

Materials:
Innertubes in a variety of sizes, building blocks, lengths of plastic pipe, air pump; large indoor or outdoor playspace

Introduction:
Place the materials, including air pump, in the large playspace and bring the children in as a group. Briefly review two safety rules: they are to ask you for help with the air pump; they are not to hit with or throw any of the materials. Then tell the children they may play with the materials as they want. Little teacher intervention should be required at first.

Continuation:
As the children begin to verbalize about their play with the innertubes, ask questions to further their explorations. Use vocabulary as appropriate. Also, encourage them to show each other fun ways to use the innertubes.

Application:
What are all the different ways you can use the innertubes?
How can you use the other materials?
What will happen if you *(name of action)*?
Please, show us how you did that *(indicate result)*.
How can you use the innertubes to *(name of result)*?

Analysis:
Why did the tube move that way?
How did you make that happen?
How is an innertube like *(different from)* a ball?
Why do you think you can use the innertube this way?
What made the block *(plastic pipe)* do that *(indicate action)*?

Synthesis:
What kind of games can you play with these things?
What could you make using all of these things?

Evaluation:
Did you enjoy playing with the innertubes?
What is the most fun?
What is the hardest thing you tried to do?

Closure:
Remind the children they will have to stop playing in about two minutes. When the two minutes are up, ask the children to roll their innertubes into one pile. Give one child a large empty box to play garbage collector. The other children should place the smaller materials into the ''garbage truck.'' As the children complete the clean-up, have them sit in a circle. Ask if anyone wants to tell about what they did with the innertubes. Use verbal elaborations and extensions as the children attempt to describe their actions. Refer to questions above to facilitate this discussion.

At the end of the discussion, ask how they would like to use the innertubes next time they play with them. These suggestions should be written down on an experience chart for future reference.

Learning check:
Introduction: Observe for following rules, exploration of materials
Continuation: Observe for demonstration of solutions, following rules; listen for responses to questions, comments on observations
Closure: Observe for following directions; listen for explanations of problems and solutions

SAND

physics; n-k   designed by Susan Gayle Golden

Purpose:
To investigate the properties of sand as it is affected by various substances

Thinking skills:
Science problem solving: Observation, hypothesis formulation, hypothesis testing
Creativity:
    Fluency: Production of alternative trials
    Flexibility: Variety of alternative trials
Communication: Verbal, non-verbal, interpersonal

Vocabulary:
Consistency, sift, pour, sticky, slimy, gooey, heavier, dry, damp, pack (verb), coarse, fine, dissolve

Prerequisites:
Extended experiences with general sandbox play

Materials:
Sand; plastic toys, such as cars, trucks, people; shovels, buckets, funnels, sifters, plastic bottles, pieces of window screen, cloth, water, paper cups, cans, sugar, salt, pebbles, paper, pencil; sandbox or sand table

Introduction:
Gather the group and review rules, including the consequences, for safe sand play. Bring the children to the sandbox or table, where the materials are already available. Allow the children to play freely, asking them questions about the appearance and texture of the sand. As children begin sifting, pouring and mixing the sand, make observations about their activities and ask them to describe what they are doing. Encourage them to try both dry and wet (damp) sand. You may model these actions as well.

Continuation:
Introduce the notion of "dissolve" by asking the children if they know what the word means. Invite them to see what happens when sand, salt, sugar, pebbles are each stirred into cups of water. Also introduce the possibility of pouring the sand through things such as the sifter or screen. Ask questions to encourage their explorations.

Application:

How will the sand *(sugar, salt, pebbles)* act when you put it *(them)* in a cup of water?
How can you use these things with the sand?
How will the sand act once it is damp?
What will you have to do to separate the sand from the pebbles *(sugar, salt)*?
How much water will you need to use to dampen the sand *(dissolve the sugar, salt, sand, pebbles)*?

Analysis:

Why did the salt *(sugar)* disappear into the water?
What other things feel *(look)* like sand?
Why can you sift some materials but not others?
How are the wet and dry sand alike *(different)*?
Why can you mold things with and make prints in the damp sand?

Synthesis:

Please tell us a story about what is happening with your cars *(trucks, people). (To be used when the child is engaging in dramatic play.)*
If you were sand, what would you like to be made into?

Evaluation:

What kind of sand is it easier *(harder)* to build with?
What are the most interesting designs you can make in your sand?
What was the best way to use *(name of material)*?

Closure:

Ask the children to decide what materials they think would be best to use for cleaning up the sand area. As they decide what to do, have them try out their ideas to see who gets his work area and materials cleanest (make sure everyone comes out a winner). When clean-up is complete, gather the group and let the children tell about what they did with the sand.

Learning check:

Introduction: Observe and listen for exploration of materials
Continuation: Observe for continued exploration and experimentation, cooperation; listen for comments and observations
Closure: Observe for cooperation

SECONDARY COLORS

physics; n-k   designed by Julie Mullis

Purpose:
To blend primary pigments in order to create secondary colors

Thinking skills:
Science problem solving: Hypothesis formulation, hypothesis testing, data collection, data interpretation, communication of solutions
Creativity:
    Fluency: Production of alternative trials
    Originality: Engaging imagination
Communication: Verbal, non-verbal

Vocabulary:
Primary color, secondary color, mix, blend, orange, purple, green, brown

Prerequisites:
Knowledge of primary (and preferably secondary) colors in the environment and ability to label primary colors

Materials:
Three large jars of water; red, yellow, and blue food coloring; red, yellow and blue tempera paints; eyedroppers, spoons, egg cartons of various colors including white, clear plastic cups, paper, paint brushes, easels, coffee stirrers; newspapers
    Color each jug of water with one of the food colorings.

Introduction:
Begin by reviewing the primary colors with the group, having the children find examples of each color from around the room. Show the children the jars of colored water, eyed-roppers, egg cartons, plastic cups, and food coloring. Ask them how they could make some new colors from the three colors in the jars. Tell them they will all be able to invent their own new colors. Bring the children to a newspaper-covered workspace so they may begin experimenting.

Continuation:
Observe for a while and ask pertinent questions from below. As the children successfully blend new colors, bring out the paints and remaining materials for them to use as well. Use questioning to facilitate their thinking.

Application:
How can you use the paints and food colorings to make a new color?
How could you make a rainbow full of colors?
What will happen if you blend *(name of color)* and *(name of other color)*?
How much of each color have you used in your new color?
How will you know which colors you have already blended?

Analysis:

How did you make this color *(indicate secondary color)*?

What happens when you blend two *(three, etc.)* colors together?

What do your new colors look like?

How are they like *(different from)* your primary colors?

How are the new colors like *(different from)* colors around our classroom?

Synthesis:

How will you display your new colors?

What would a rainbow of your new colors look like?

Evaluation:

Which is your favorite new color?

What was the best way to mix new colors?

If you could be a color, which would you be?

Closure:

Give the children a two minute warning that they will need to finish with their colors. Ask them to choose a sample of their favorite new color and bring it to the group (make sure it can be displayed neatly). In the group, have them share their new colors and tell how they made them. Ask them also to talk about anything else they did while mixing colors.

When the discussion is done, the children should throw out any of the used up materials.

Learning check:

Introduction: Listen for suggested alternatives

Continuation: Observe for experimentation; listen for comments, responses to questions

Closure: Observe and listen for sharing of solutions

SIFTING

physics; n-k

Purpose:
To investigate the relationship between the size of an object and the size of the opening which accommodates it

Thinking skills:
Science problem solving: Observation, data collection, data interpretation, solution finding, solution testing, communication of solutions
Creativity:
  Elaboration: Refinement and detailing of final solutions
Communication: Non-verbal, listening, interpersonal

Vocabulary:
Sift, sort, smaller, larger, fine, coarse, medium,

Prerequisites:
Prior experience with sifters in sand play

Materials:
Assortment of sifters ranging from very fine to coarse; pieces of various grades of screening with edges rolled; chicken wire or other open screening with edges rolled; flour, sand, gravel, small rocks, rice, dried lentils, dried navy beans, dried pinto beans, dried large lima beans, plastic or wooden beads (about ½ in. in diameter), 1 in. cubes, any other items which can be sifted; newspapers or large sheets of plastic

Set out containers of the different materials and the various sifters on a large newspaper covered workspace. Mix together a number of the materials and set aside.

Introduction:
Gather the children and show them the mixed up materials. Ask how they could separate all the different things (label contents as you discuss the problem). Accept their suggestions and encourage them to think through the problem. If someone suggests sifting, hold up a very coarse screen and ask if it would be of any help. Try to elicit more specific descriptions of what is needed.

Once the children have discussed the problem for a little while, tell them you have some materials to help them solve the problem. Show them the materials, telling them they are still sorted to help the children learn more about each material. Tell the children they may work with the materials however they want to. (You may want to remind them that there is to be no throwing.)

Continuation:
As the children work they will be making piles of mixed up materials. Use questions to address the problems these piles make as well as other problems related to sifting. Use vocabulary as appropriate.

Application:
What will happen if you try to use a coarse *(medium, fine)* sifter with the *(name of material)*?
What size sifter will it take to separate these *(indicate)* materials?
How do you plan to solve our problem?
How might someone use a sifter at home?
What will happen if you sift different materials through the same sifter?

Analysis:
Why does *(doesn't)* the *(name of material)* go through your sifter?
Why will *(won't)* this *(indicate)* sifter separate the *(names of materials)*?
How does a sifter work?
What usually happens when you try to separate coarse *(medium, fine)* materials?
What usually happens when you use a coarse *(medium, fine)* sifter?

Synthesis:
How would you design a machine to sift out all these materials at one time
If we wanted to sift real cars from real houses, what would the sifter have to look like?

Evaluation:
Which sifter works best for sifting out the (name of material)?
What will be the best way to solve our problem?

Closure:
Go around to each child and identify one material for him or her to sift out. Ask that as much of this material as possible be returned to the original container. Remind the children to find the best sifters for the jobs they have been assigned. When they are done cleaning up what is on the table, ask them to try to solve the problem of the container of mixed up materials. Give the children a chance to do some of the sifting. When completely done sifting, the children should help with the rest of clean-up.

Learning check:
Introduction: Observe and listen for participation in discussion, exploration of materials, following instructions
Continuation: Observe and listen for attempts at solutions, cooperation, responses to questions
Closure: Observe and listen for successful participation in clean-up, cooperation, sharing

SOLIDS AND LIQUIDS

chemistry; n-k

Purpose:
To discover the various properties of liquids and solids

Thinking skills:
Science problem solving: Observation, hypothesis formulation, hypothesis testing, communication of solutions
Creativity:
   Fluency: Production of alternative trials
   Flexibility: Variety of alternative trials
Communication: Verbal

Vocabulary:
Solid, liquid, pour, shape, container, same, different, hard, soft, scratch

Prerequisites:
Prior experience with pouring activities

Materials:
Liquids: water, corn syrup, vinegar, rubbing alcohol, vegetable oil, honey, milk, tempera paint
Solids: wood blocks and beads, pieces of rubber and solidplastic, rice, dried beans or peas, rocks, chunks of metal *(Note: Do not use powders.)*
Ample paper cups, funnels, spoons, small hammers, dish pans *(one filled with warm water)* surface to bang on (like a sidewalk), smocks, newspapers, old towels
   Set up the work area on the newspapers with dish pans to pour over (into).

Introduction:
Put smocks on the children, put them in groups of 3-4, and show them the liquids and solids. Tell them you have a problem: all these materials need to be put into two groups so that everything in a group is alike in some way. Remind them not to taste anything. Show them the remaining materials and tell the children that these are things they can use to help with their decisions. Observe as they begin, showing the children how to place objects inside the towel before hitting with hammers.

Continuation:
Continue to observe, assisting as needed. Ask questions and introduce vocabulary to provoke deeper investigation.

Application:
How will you use the tools to help you?
What will happen when you hammer on the *(name of material)*?
What will happen if you try to stir the *(name of material)*?
What kinds of things do we use liquids *(solids)* for?
What will happen when you put something solid into something liquid?

Analysis:
How are *(name of material)* and *(name of other material)* alike *(different)*?
How is the rice *(beans, peas)* like a solid *(liquid)*?
Why does *(doesn't)* *(name of material)* pour?
What can you say about all of the solids *(liquids)*?
Are you a solid or a liquid?

Synthesis:
What would something be like if it were both a solid and a liquid at the same time?
If you were liquid, how would you move?

Evaluation:
Which materials are easier to contain?
Would you rather play with solids or liquids?

Closure:
Give the children a two minute notice. During this time, have each group sort its materials, first removing tools then putting remaining materials into two groups. As they sort, ask them to defend their two groups, especially as to their placement of the rice, beans and/or peas. Also ask for characteristics of liquids and solids. Once they have shared their sorting with you, ask that all the rocks be put in one container, rice in another, etc., until every possible material is cleared away. Then they should pour their liquids into a plastic garbage bag and wipe out the dish pans.

Learning check:
Introduction: Observe and listen for exploration of materials and related comments
Continuation: Observe and listen for experimentation with materials and relevant comments on observations; listen for responses to questions
Closure: Observe for appropriate classification; listen for correct explanations

TARGET BALL

physics; n-k   designed by Julie Lynn Key

Purpose:
To knock over targets using balls of a variety of sizes, weights, and shapes

Thinking skills:
Science problem solving: Observation, solution finding, solution testing, communication
of solutions
Creativity:
   Fluency: Production of alternative solutions
Communication: Verbal, non-verbal, listening

Vocabulary:
Size, weight, shape, target, line, stack, heavy, light, larger, smaller, same, different

Prerequisites:
Prior play with all types of balls

Materials:
Orange juice cans or the like; several sizes of footballs, basketballs, beach balls, tennis
balls, golf balls, ping pong balls, 10-12'' hollow plastic balls, nerf balls; large floor space

Introduction:
Ask the children as a group if any of them have ever bowled before. Discuss what they
know and remind them that in bowling the ball always stays on the floor. Tell the children
they may take what they want and try ''bowling'' or target ball with the cans and balls.

Continuation:
During this free play, walk around and talk to the children about what they are doing.
Encourage them to use the materials in challenging ways by creating different arrangements
of targets, making comparisons of the different kinds of balls, and the like. Use questions
to promote high level thinking.

Application:
How will you use these materials to make targets?
Which ball will be easiest *(hardest)* to aim?
How many cans do you think you can hit in one roll of the ball?
In what other games do you throw or roll a ball to hit something?
How will you arrange your target?

Analysis:
How did the different balls work for hitting your target?
Why did you knock over only part of your target?
What happened when you used a large *(small, light, heavy)* ball?
What happens when you spread out *(stack up, push together)* the pieces of your target?
How is your target ball game like *(different from) (name of other child)*?

Synthesis:
What other materials should we add to target ball the next time we play?
What rules will improve your target ball game?

Evaluation:
Which target arrangement *(ball)* worked best *(worst)*?
Would you like to play this kind of game again using different materials?

Closure:
Give a brief warning before clean-up. Ask the children to come to the circle, each with the ball he thinks works best. Have them each show the balls, and try to say why they are best. Encourage questions and responses from the children. After discussion, the children should then put away their materials in designated containers. It would be helpful if the clean-up time were a game as well.

Learning check:
Introduction: Listen for comments about prior experiences
Continuation: Observe for experimentation with materials, demonstrations of successful solutions; listen for comments about experiences and observations, responses to questions
Closure: Observe for participation; listen for explanations

TRACKS

designed by Luwonna Ellis

Purpose:
To create and differentiate among tracks made by various objects in dry and wet sand

Thinking skills:
Science problem solving: Problem finding, observation, hypothesis formulation, hypothesis testing
Creativity:
   Fluency: Production of alternative trials
   Flexibility: Variety of alternative trials; uniqueness of proposed solutions
Communication: Verbal, non-verbal

Vocabulary:
(Reinforcement) rough, smooth, damp, wet, dry, cool, push, pull, drag, slide, roll, track

Prerequisites:
Experience playing with wheel toys and with wet and dry sand

Materials:
Dry sand, wet sand, water; sandbox tools, wheel toys, balls, marbles, wooden dowels, small cans; sandbox or sand table

Introduction:
Set out the materials where they are accessible to the children. Invite them to explore what you have set out, requesting that they keep the dry sand dry. Some of them will likely engage almost immediately in fantasy play, so ask them what else they can do with the materials. Also be sure to comment on any tracks you observe a child making with the objects.

Continuation:
Continue making comments about the tracks the objects make, asking pertinent questions. Reinforce correct use of vocabulary. You may need to suggest or model different things the children can do with the objects.

Application:
How can you use these objects in the sand?
What kind of track do you expect the *(name of object)* to make?
How hard do you have to push to make a track with the *(name of object)*?
What object probably made this *(point to)* track?
How can we use the tracks we see?

Analysis:
How are the wet and dry tracks alike *(different)?*
What happens when you push *(pull, drag, slide, roll)* the *(name of object)* through the wet *(dry)* sand?
How are these two objects alike *(different)?*
How are the tracks like the objects that make them?
How does it feel when you roll *(push, pull, drag, slide) (name of object)* in the wet *(dry)* sand?

Synthesis:
How can you combine your objects to make a new and unusual track?
How can you make a picture with tracks in the sand?

Evaluation:
Which tracks are easier to identify?
Which tracks tell us most about the objects that make them?

Closure:
Ask the children to make one last track from an object where they are working and then put all the objects aside. Ask them to look carefully at the track and then come sit in a circle for a few minutes. When they get into the circle ask for descriptions of their last tracks. Ask what other tracks they made and how they looked. Also, ask them to share anything special they noticed when making the tracks.

Learning check:
Introduction: Observe for exploration of materials
Continuation: Observe and listen for experimentation with materials; listen for observations and suggestions; observe for demonstrations of discoveries
Closure: Observe for construction of solutions; listen for observations, comments

WATER PLAY WITH CONTAINERS

physics; n-k

Purpose:
To investigate the effects of water pressure and gravity on water in containers

Thinking skills:
Science problem solving: Problem finding, observation, hypothesis formulation, hypothesis testing
Creativity:
  Fluency: Production of alternative trials
  Flexibility: Variety of alternative trials
Communication: Verbal, non-verbal

Vocabulary:
Container, perforate, perforated

Prerequisites:
Prior experiences with other types of water play

Materials:
Tubs of water; cans and plastic containers of various sizes, some with snug-fitting lids and one or several holes in the bottom, some without lids and one or several holes in the bottom, some with bottoms intact (*Note: in cans, perforations should be smoothed for safe handling*); smocks, towels, newspapers, mops

Introduction:
Ask the children to put on their smocks and then put the children in groups of 3-4 per tub of water. Tell them they may play with the materials however they wish but to try to keep the water in the tubs. Little intervention should be necessary during this free exploration.

Continuation:
Observe how the children are working and comment on their actions. To encourage extended exploration and experimentation, ask provocative questions, introducing vocabulary as appropriate.

Application:

What do you expect to happen when you push down on this *(indicate)* container?

What will the water do when you lift up this can *(container)*?

How will it be different when you put the lid on?

What are you going to do with your materials?

How do people use containers with holes in them?

Analysis:

Why did the container float *(sink)*?

How do the containers with holes *(solid bottoms, lids, one hole)* usually act?

How are these *(indicate)* containers similar *(different)*?

What made the water behave like this?

How will you decide if a new container will sink or float?

Synthesis:

Develop a new use for containers with holes.

How would you design a water play toy using some of these containers?

Evaluation:

Which kinds of containers make the best showers?

What did you enjoy most about water play today?

Closure:

Give the children a two minute warning before asking that they place all the containers in an empty tub and dump out the excess water. You should also involve them in any wiping or mopping up which may be required. When clean-up is complete, the children should remove smocks and gather for a discussion of the activity. Focus on what they discovered about water and the containers and ask them to plan for what other materials they would like to use with the containers. Keep a list of their suggestions for future reference.

Learning check:

Introduction: Observe and listen for exploration of materials, observations

Continuation: Observe for experimentation with materials; observe and listen for responses to questions

Closure: Observe and listen for participation in clean-up, contributions to final discussion

WETTABLE MATERIALS

physics; n-k   designed by Molly Suzanna Foster

Purpose:
To categorize materials by how they respond to being exposed to water

Thinking skills:
Science problem solving: Observation, data collection, data interpretation
Creativity:
   Fluency: Production of alternative trials
Communication: Verbal

Vocabulary:
Absorb, dissolve, texture, separate (verb)

Prerequisites:
Water play and sink/float experiences

Materials:
Buckets or dishpans half full of warm water, cups of warm water; sand, salt, sugar, cornmeal, rice, soil, corks, nails, cotton balls, bottle caps, pebbles, wood chips, waxed paper, colored paper; small bowls or cups, spoons, sponges, sifters; paper towels for clean-up, smocks, newspapers

Set out pans of water, corks, nails, cotton balls, bottle caps, pebbles, wood chips, pieces of waxed and colored paper, sponges at one table which has been covered with newspapers. At a separate table covered with newspapers, set out sand, salt, sugar, cornmeal, rice, soil, bowls or cups, and sifters.

Introduction:
Have the children all put on their smocks. Tell them they are each going to have a turn working with the water and with the sifters. Split the group in half and ask one group to go to the sifters and see how easily the different materials will go through a sifter. Ask the other half to go to the water and figure out what happens to their materials when they get wet.

Continuation:
Circulate between the two tables and observe how the children are interacting with the materials. At the sifting table, note when the children have tried sifting most or all of the different materials. At this time, introduce cups of warm water and ask the children what will happen in the sifters if their materials are wet. Tell them they may use the cups to mix things in.

Begin asking each group pertinent questions and using appropriate vocabulary.

Application:
What will happen to the *(name of object/substance)* when it gets wet?
What will *(name of material)* be like when it dries out?
How can you get the *(name of substance)* to go through the sifter?
How can you separate the water from the *(name of substance)*?
What are some uses for wet *(name of substance)*?

Analysis:
Why·do you think this *(indicate specific reaction)* happened?
How are the wet materials alike *(different)*?
How have the textures of *(name materials/substances)* changed since you wet them?
Why does the water make the *(name of substance)* stick *(lump)* together?
What happened to the salt *(sugar)*?

Synthesis:
What will your sifter need to be like to sift the wet *(name of substance)*?
What will you get if you mix all these materials together?

Evaluation:
Are the substances more different or more alike now that they are wet *(dry again)*?
Which materials got the wettest?
Which substance was *(will be)* easiest *(hardest)* to sift *(wet, dry out)*?

After about 15 minutes, switch the two groups so each group has ample opportunity to work with the water and with the sifters, being sure to start the second groups with fresh materials. (This activity could be carried out over a two-day period as well as all on one day.)

Closure:
Give the children a two-minute warning before clean-up. At the end of two minutes give one child a garbage can and another an empty bucket. Ask that the wet substances by the sifters be put in the garbage can and the wet materials by the water be put in the bucket. Have the rest of the children work together to empty the water, rinse the sifters, pick up leftover dry substances, and throw out used newspapers. They should also sponge clean the table tops. Then they can dry hands and take off their smocks.

When they are cleaned up and dried off, get the group together to talk about the activity. Spend about five minutes, asking what they did and reinforcing concepts and vocabulary.

Learning check:
Introduction: Observe for exploration of materials
Continuation: Listen for comments about observations, responses to questions; observe for experimentation with materials
Closure: Listen for descriptions of observations and conclusions

COOKING

chemistry; n-k/primary

Purpose:
To experiment with simple, quick-cook foods

Thinking skills:
Science problem solving: Observation, solution finding, solution testing, communication of solutions
Creativity:
   Originality: Uniqueness of solutions
   Elaboration: Detailing of solutions
Communication: Verbal, interpersonal

Vocabulary:
Mixture, mix, stir, thicken, heat, chill, boil, simmer, set (verb)

Prerequisites:
None

Materials:
Instant and cooked pudding mixes, jello mixes, instant soft drink, tea, cocoa mixes, dry powdered milk; water, soda, liquid milk; bowls, spoons, bowl scrapers, plastic cups, sauce pans, stove or hot plate, refrigerator; paper towels, sponges.
   Set out the materials so they are clearly identifiable.

Introduction:
Show the children the dry foods, labeling the substances as you do. Tell them these are all foods that people mix to eat or drink and that you would like them to make up their own recipes with these foods. Organize groups of 3-4 and let them explore, discuss and begin experimenting. Allow them to taste a tiny bit of a substance if they need to.

Continuation:
Observe the children's progress and begin asking questions to further their thinking. Introduce vocabulary throughout discussions.

Application:
How do you plan to use your ingredients?
What will happen when you mix *(names of two substances)*?
What are you planning to make?
What do you expect your mixture to taste *(look, smell, feel)* like?
What will happen if you heat *(chill)* your mixture?

Analysis:
Why does your mixture taste *(look, smell, feel)* like it does?
What happened when you mixed *(names of two substances)*?
What usually happens when you mix a liquid and a powder?
Why did your mixture thicken *(set)*?
How are these mixtures similar *(different)*?

Synthesis:
What new food are you designing?
What new ingredients would you like to use to improve your mixture?

Evaluation:
Which mixture do you like best?
Is taste, smell, appearance, or texture most important in picking which is best?

Closure:
As the children complete their cooking, set products aside and have them clean up. Make sure they wash their dishes if facilities are available. Once each group has cleaned up, let the children help themselves to tastes of the various mixtures. (No one should get more than a small dab of each or be expected to eat it all.) Permit spontaneous discussions, but try to refocus children who start making derogatory remarks.

Learning check:
Introduction: Observe and listen for cooperation, beginning exploration of materials and related comments
Continuation: Observe and listen for continued cooperation, progress toward a successful solution, responses to questions
Closure: Observe for participation in clean-up, sharing of solutions

HEATING AND COOLING

physics; n-k/primary   designed by Johnna K. Bolick

Purpose:
To cause changes in the states of various kinds of matter

Thinking skills:
Science problem solving: Observation, hypothesis formulation, hypothesis testing, solution finding, solution testing, communication of solutions
Creativity:
  Originality: Uniqueness of combinations
  Elaboration: Detailing of solutions
Communication: Verbal

Vocabulary:
Heat, cool, melt, solidify, temperature

Prerequisites:
Knowledge of solid and liquid forms of matter

Materials:
Hot plates, refrigerator/freezer; bowls, spoons, pots, potholders; ice, chocolate, paraffin, butter, jello (already made up), rocks, small pieces of wood, marbles, salt, sugar, flour, water, rubbing alcohol

Introduction:
Pour out a container of ice and ask the children what will happen to the ice now that it is out of the freezer. Hand around pieces of ice for them to feel and watch, while they review solids and liquids. Ask questions about why the ice will melt and about the differences between solids and liquids. Have the children explain what they experience as the water changes from solid to liquid. Also, ask how it can be made solid again. Be sure to try out the children's suggestions.

Continuation:
Bring out the other substances and ask the children for their suggestions to change the states of the substances. Keep a chart of their ideas and then let them try to effect the changes as they suggested. Use questions to further their experimentation and understanding and introduce the vocabulary as appropriate.

Application:
How can you make *(name of substance)* become a liquid *(solid)*?
What could you do to make *(name of substance)* melt *(solidify)* faster?
What will happen to the liquids if you add salt *(sugar, flour, water, or any of the other substances)*?
How do people use substances which change from solids to liquids *(liquids to solids)*?
What will happen when we heat *(chill)* the *(name of substance)*?

Analysis:
How are the solid and liquid states of *(name of substance)* similar *(different)*?
Why do you think *(name of substance)* is *(not)* changing form?
What can you generally say about some of your solids *(liquids)*?
Which methods of changing substances work, and why?
Why do changes in temperature affect the different substances in different ways?

Synthesis:
How would you combine several substances to make a new substance which will melt *(solidify)* very easily?
What would you feel *(look)* like if you melted?

Evaluation:
Which methods for changing substances worked best *(least well)*?
Which substance was easiest to change in form?
Would you rather be a solid or a liquid?

Closure:
Give the children a five minute time warning and have them clean up their work areas after the five minutes have elapsed. If they have put things in the refrigerator or freezer, tell them to go observe what is happening and then come back to the group. Once all the children are back together, discuss what they observed. Also check on and talk about their experiments with recreating the ice from the beginning of the lesson. Focus especially on how temperature can change the physical state of a certain substance.

Learning check:
Introduction: Listen for comments, possible explanations; observe for exploration of materials
Continuation: Listen and observe for exploration of and experimentation with materials, refinement of ideas
Closure: Listen for sharing of ideas and experiences

INCLINE PLANES AND TARGETS

physics; n-k/primary

designed by Julie Lynn Key

Purpose:

To experiment with how the angle of an incline plane affects speed and path of various objects rolled down it

Thinking skills:

Science problem solving: Problem finding, observation, hypothesis formulation, hypothesis testing

Creativity:

Fluency: Production of alternative trials

Flexibility: Variety of alternative trials and uses

Originality: Uniqueness of solutions

Elaboration: detailing of hypotheses and solutions

Communication: Verbal, interpersonal

Vocabulary:

Incline plane, target, angle, slant, ramp, slope

Prerequisites:

Prior experience playing with a variety of objects and wheeled toys on incline planes

Materials:

Boards of several lengths and widths; balls and marbles, including several shapes and sizes; empty cartons, cans, plastic jugs; several building blocks, including wedges

Introduction:

Set out the materials in a convenient location and tell the children they may select items of interest to experiment with. Ask that nothing be thrown but say that things can be rolled. As the children begin free exploration, observe how they use the materials and encourage the rest of the group to observe a child involved in an interesting experiment. Begin introducing relevant vocabulary through your comments and observations. Be sure that this initial period of 5-10 minutes is allotted to unstructured free exploration of the materials.

Continuation:

After about 5-10 minutes, begin interacting with the children about what they are doing. Observe and talk with them about how they are using their materials. Ask questions to guide them into constructing ramps and targets. Encourage them to experiment with tilting the ramps sideways and creating ramps that incorporate turns or bends. Refer to the questions below and others as appropriate, using relevant vocabulary during discussions.

Application:
How can you use these different materials?
What are some of the ways incline planes (*ramps*) are useful to people?
What will happen if the incline plane tilts sideways, too?
What would happen if you made something similar to a sliding board and rolled a ball down it?
What can you do with all these other objects?

Analysis:
What do you need to do to hit the target?
What usually happens when you tilt the ramp sideways?
How are the various balls alike (*different*) in their behaviors on the incline planes?
How does the slope (*length*) of the incline plane affect the roll of the ball?
What happens when an incline plane turns a corner?

Synthesis:
How would you design a better incline plane (*ramp*)?
How can you organize your information so that it helps other people learn about incline planes?
What kind of new ballgame could you make using incline planes and targets?

Evaluation:
Which was your best (*worst*) attempt at hitting a target?
What was the best way to get an incline plane to turn a corner?
What is the most important use of the incline plane?

Closure:
After the children have had sufficient time to experiment, go around and tell each group that they need to finish in a few minutes. Then ask the children to come together to share their experiences and discuss whatever is concerning them about their experiments. Encourage comparison of findings by asking more open-ended questions. When the discussion is over, ask each of the children to help clean up the work space.

Learning check:
Introduction: Observe for interactions with peers and materials; listen for observations, ideas, suggested explanations
Continuation: Observe for extent of experimentation, social interactions; listen for exchanges of ideas, extent of communications
Closure: Observe and listen for substantive participation in discussion, cooperation

PATTERNS

physics; n-k/primary   designed by Michael Lee Buie & Julie Mullis

Purpose:
To explore how different patterns are formed by moving objects

Thinking skills:
Science problem solving: Observation, data collection, data organization, data interpretation
Creativity:
    Originality: Uniqueness of solutions
    Elaboration: Detailing of solutions
Communication: Verbal, non-verbal

Vocabulary:
Pattern, track, design, curve, line, merge, straight

Prerequisites:
Experience making patterns using things like objects and ink pads, or hand and foot prints

Materials:
Cake pans and dish pans of various shapes and sizes, ample supply of white (and colored) paper cut to fit pans; thinned tempera paints in low, flat containers, spoons, buckets of clean water; marbles, blocks, balls, dried beans, pencils, thread spools, golf tees, beads of various shapes and sizes, crayon pieces, bottle caps, dried peach pits, nuts, seeds, seashells, nails, screws, nuts and bolts, small cubes; smocks, newspapers
    Set out the materials at a newspaper covered table.

Introduction:
Put on smocks. Spend a few minutes experimenting with the children to see whether a selection of the materials roll or slide across the pans lined with paper. Ask the children how they can tell where an object has moved across the pan. Then explain to them that they will be able to trace the paths of their objects. Together, each dip an object into the paint (model using a spoon to do this) and then set it in the pan.
    Encourage the children to get the objects to move in their pans and see what happens. As they experiment, talk about their observations, emphasizing the patterns (tracks) made.

Continuation:
Offer them the entire array of materials to experiment with. When they are done with an object, it belongs in the bucket of water (you will have to wash and dry). As their papers become dense with tracks, offer fresh sheets. If someone wants to save what s/he's done, set it aside with the child's name on it. As the children work, make observations about what they are doing and ask questions as appropriate.

Application:
What will happen if you use more than one color?
What kind of pattern *(track, design)* do you expect a *(name of object)* to make?
What do you need to do to make the patterns *(tracks, designs)* change?
What will happen when two patterns *(tracks, designs)* cross each other *(merge)*?
How will you keep a record of which pattern *(track, design)* goes with which object?

Analysis:
Which materials make very similar *(different)* patterns *(tracks, designs)*?
How are these objects *(patterns)* alike *(different)*?
What usually happens when you use more than one color?
How come the *(name of object)* makes this *(indicate)* kind of pattern *(track, design)*?
How is this pattern *(track, design)* like *(different from)* the object that made it?

Synthesis:
What kinds of unique pictures can you make with various patterns *(tracks, designs)*?
How would you design an object to make an unusual pattern?

Evaluation:
Which of your objects made the most interesting *(unusual)* pattern *(track, design)*?
How well did you enjoy creating patterns?
Which of your pattern sheets will you decide to keep?

Closure:
Give the children a two to five minute warning for clean-up. When time is up, ask them to roll all of their objects into the buckets of water. Each child should then remove the paper, wipe his pan well with paper towels and set the pan aside.

Then he should remove the newspaper from his workspace. When hands are washed and smocks are off, ask the children which shapes they enjoyed the most. Explain that they will be able to do this again in a few days. Discuss finding new objects to bring from home and what they would prefer these objects to be like.

Learning check:
Introduction: Observe and listen for participation and exploration
Continuation: Observe and listen for accumulation of information from observations, successful solutions, responses to questions
Closure: Listen for extensions of ideas

PENDULUMS WITH TARGETS

physics; n-k/primary   designed by Dana L. Walker

Purpose:
To control the construction of the target and the swing of the pendulum so that the target can be hit

Thinking skills:
Science problem solving: Observing, hypothesis formulation, hypothesis testing, solution finding
Creativity:
  Fluency: Production of alternative trials
  Flexibility: Variety of alternative trials
Communication: Listening, interpersonal

Vocabulary:
Pendulum, bob, swing (noun), propel, circular, direction

Prerequisites:
Experience with constructing pendulums and varying the length of string, size/weight of bob, and release

Materials:
Long pieces of string, pieces of wood with screw eyes in one end; tin cans, building blocks, stuffed animals, plastic 1 and 2 liter bottles, egg cartons, plastic milk jugs; ceiling from which pendulums can be hung or structures to make pendulums taller than the children

Introduction:
Have a model pendulum hanging somewhere in the room and put the children in groups of 3-5, depending on age. Show the children the pendulum and ask them to try to make one like it, giving them the needed materials. Help them with knots and with hanging the pendulums overhead. Discuss what they remember about pendulums from their prior experiences and let them swing the pendulums a bit to get the feel of them. Meanwhile, be getting out the remaining materials. Ask the children how they could use the pendulums to knock over targets.

Continuation:
Permit the children to select their own materials for targets and experiment with the relationship between the target and the pendulum. As they work, ask questions which further their thinking.

Application:
How will the pendulum act if you push *(pull, drop, swing)* it?
How can you use the materials to make a tall *(short, narrow, wide, big, noisy, quiet, etc.)* target?
How will the pendulum act if you stand in a different place when you let it go?
What will happen if you move the target to here *(indicate location)*?
Which materials will make the best *(worst)* target?

Analysis:

Why did the pendulum miss *(hit)* the target?

How is it different when you push and when you drop the bob?

What happens when you push *(drop)* the bob?

How are the target materials alike *(different)*?

What usually happens when the bob hits a tall *(short, heavy, light, wide, narrow)* target?

Synthesis:

How would you make a very accurate pendulum?

How will you decide what is the best possible pendulum and target?

What kinds of games can we play with pendulums and targets?

Evaluation:

Which target was easiest *(hardest)* to hit?

What is the best way to get a pendulum to swing where you want it to?

Which target made the most interesting sounds?

Closure:

Tell the children they have time to make one more target and that this time they should use all of their materials in the biggest target they can build. Then see how many swings of the pendulum it takes to knock the targets down. As the targets are knocked down, the children should begin putting their materials in a storage box or the like. Ask them to assist you in taking down the pendulums. Once everything is put away, regroup for a brief review of what they discovered during the activity.

Learning check:

Introduction: Observe and listen for attempts to copy the model, cooperation, notation of details

Continuation: Observe for experimentation, continued cooperation; listen for idea sharing

Closure: Observe and listen for collaboration on final solution

SIMPLE PULLEYS

physics; n-k/primary

Purpose:
To construct and use simple pulleys for moving objects horizontally and vertically

Thinking skills:
Science problem solving: Observation, solution finding, solution testing, communication
   of solutions
Creativity:
   Originality: Unusualness of solutions
   Elaboration: Refinement of solutions
Communication: Verbal, interpersonal

Vocabulary:
Pulley, line, weight, resistance, gravity, horizontal, vertical

Prerequisites:
None

Materials:
Pulleys of various sizes, at least 2 alike in some sizes and including several pairs of washline pulleys; long pieces of washline; buckets, empty jugs, blocks and other objects to be carried in buckets, clothespins, paper, pencils, "S" hooks (can be made out of bent hanger wire), doll or dress-up clothes; climbing frame or horizontal bars for vertical pulleys; cookies or other individual snacks

   Set up two or three clotheslines with pulleys at each end. Array other materials and remaining pulleys in a convenient location.

Introduction:
Gather the children at one end of a washline. You go to the other end and send them a package of the snacks across on the line using the pulleys. Ask that someone send the empty container back to you on the line. Tell the children that once they are done eating the snack, they may begin to send each other things on the lines and pulleys or experiment with the other materials. Try not to interact at first, unless your help is requested.

Continuation:
Observe the children working and see how well they are employing other materials. If no one has set up a vertical pulley system, set one up and model using it to lift a bucket. Use the vocabulary as you comment on the children's activities and begin asking questions to stimulate further experimentation.

Application:
What are some ways you can use these materials?
How do you plan to get the *(name of object)* from one side to the other?
What do you expect to happen when you pull on the rope?
How do people use pulleys?
What will you have to do with the rope to get your pulley system to work?

Analysis:
How come the *(name of object)* moved in that direction?
How is the vertical pulley helping you lift the *(name of object)*?
What usually happens when two people pull the line at the same time?
How does the horizontal *(vertical)* pulley work?
How are the vertical and horizontal pulleys alike *(different)*?

Synthesis:
How would you design a system using 3 *(or other larger number)* pulleys to carry things around our classroom?
What kind of games could we play using pulley systems?

Evaluation:
Was the vertical pulley or the horizontal pulley more helpful?
What is the most important way people use pulleys?

Closure:
Give the children a two minute warning to finish whatever they are doing. Then bring them together and ask them to share what they tried to do with the pulleys. Encourage them to be as specific as possible.

After the discussion, everyone should help clean up. You will probably need to assist with the washlines.

Learning check:
Introduction: Observe and listen for observations, exploration of materials, cooperation
Continuation: Observe for successful uses of pulleys; listen for responses to questions, collaboration among peers
Closure: Listen for successful participation in discussion; observe for cooperation during clean-up

SUCTION

physics; n-k/primary

Purpose:
To investigate how sucking actions and suction function to move objects

Thinking skills:
Science problem solving: Problem finding, observation, solution finding, solution testing, communication of solutions
Creativity:
  Elaboration: Refinement of solutions
Communication: Verbal, non-verbal

Vocabulary:
Suck, suction, light, heavy, lift, inhale, exhale, porous

Prerequisites:
Prior experience with blowing activities and with drinking through straws

Materials:
Scraps of paper, facial tissues, full-sized sheets of paper, green leaves, small (1 in. x 1 in. or a bit larger) pieces of flat plastic and aluminum foil (but NOT plastic wrap), bottle caps, paper clips, several large sheets of paper, small paper or plastic cups, buttons (½ in. or larger), wadded up scraps of paper, scraps of fabric of various kinds, pebbles (½ in. or larger), dried lima beans; several straws per child, empty buckets

Set materials out on a table, reserving some items for later. Several yards away, place buckets on the floor or on another table.

Introduction:
Show the children the materials, except straws, and ask them how they might move these materials into the buckets without touching the materials with any parts of their bodies. Accept suggestions and then show the straws and ask how the straws might help. Accept more suggestions and then have the children try out their ideas.

Continuation:
As the children are working, observe how they are responding to the different materials. Ask questions and introduce vocabulary as appropriate.

Application:
How will you use the straws?
What will happen if you try to use several straws at once?
How do people use suction to move objects?
What will happen if you try to blow *(carry)* things from here to the buckets?
How can several of you work together to solve the problem?

Analysis:

Why can *(can't)* you pick up the *(name of object)* with your straw?
What usually happens when you try to pick up a porous *(large, smooth, heavy, lumpy, light)* item?
How are the items that you can move alike *(different)*?
Why does suction work on only some of the objects?
What happens when you try to use several straws at one time?

Synthesis:

How would you build a suction machine for lifting cars *(or some other massive object)*?
How can all of you work together to move the large sheet of paper?

Evaluation:

Which materials are easiest *(hardest)* to move?
What conditions made moving the objects easiest *(most difficult)*?

Closure:

Lay out the remaining materials and ask the children to find any way they can to get all of the objects into the buckets without ever touching the objects or buckets. If your children cooperate well, you could set this up as some type of contest (without prizes). Once all of the materials are in the buckets, ask everyone to sit down, relax, and take several SLOW deep breaths. Then ask them to review briefly what they discovered.

Learning check:

Introduction: Listen for suggestions; observe and listen for beginning of experimentation with materials and related comments
Continuation: Listen and observe for continued experimentation and refinements of actions
Closure: Listen and observe for successful participation in final activity and discussion

TEXTURED PAINT

physics and chemistry; n-k/primary   designed by Martha A. Sweeney

Purpose:
To explore how different materials change the texture and consistency of paint

Thinking skills:
Science problem solving: Observation, problem finding, hypothesis formulation, hypothesis testing
Creativity:
   Originality: Engaging in playful imagination, uniqueness of alternative trials and uses;
   Elaboration: Detailing of alternative trials and uses
Communication: Verbal

Vocabulary:
Texture, consistency, mixture, descriptive words like lumpy, slimy, gritty, grainy, mushy, watery

Prerequisites:
Familiarity with tempera paints

Materials:
Large containers of prepared tempera paints; newspapers, paper or plastic cups, plastic spoons, plain paper, large tables; four groups of additional materials: group 1 - vinegar, oil, baking soda; group 2 - jello (solidified), pudding (solidified), cooked oatmeal, cooked cream of wheat or grits, applesauce; group 3 - rock salt, sand, sugar, flour, Grape Nuts cereal, popped corn; group 4 - cotton balls, toilet paper, yarn, string

Introduction:
Arrange children so they are working in four groups at newspaper covered tables. Provide each group with plain paints and paper and invite them to paint using their fingers. As they begin, add a group of materials to each table, asking the children to think about how they can use these new materials with their paints. Be sure to tell them not to taste anything.

Continuation:
Follow the leads of the children as they explore different combinations of materials. Encourage them to paint with their textured colors to see the visual effects. As you observe them, ask questions and encourage the use of descriptive vocabulary.

Application:
What can you think of to do with these materials?
What will happen if you mix *(name of substance)* with the paint?
How can you use a textured paint when you make a painting?
What will you have to do to make this paint thicker *(thinner, smoother, rougher, etc.)*?
How much *(name of substance)* will it take to make the paint feel much different?

Analysis:

Why did the paint change in this way?

How are these paints alike *(different)*?

What happened to the rock salt *(sugar)*?

Why can you feel *(name of substance)* but not *(name of other substance)* once they are each mixed into paints?

What happened when you tried to paint with this mixture?

Synthesis:

What kind of painting could you make using all these different textures and colors?

What words can you invent to name the new textured paints?

If you saw this texture on a painting, what would it remind you of?

Evaluation:

Which paint has the most *(least)* interesting texture?

Which paint will be the hardest *(easiest)* to use in painting a picture?

Were you able to get the textures you wanted?

Closure:

After sufficient time for experimentation, quietly assign each child to a clean-up task as you warn him/her to finish the experiment. (You could hand out slips of paper with clean-up tasks pictured on them.) Ask the children to set aside any paintings they wish to keep. Then ask that they clean up according to their assigned tasks. (Be sure to have a large plastic garbage bag for excess paints as they will clog the sink drain. Or cover them tightly for use on another day.) As the children clean up, talk to them individually or in small groups, asking questions to ascertain what they have learned.

Learning check:

Introduction: Listen for comments and suggestions

Continuation: Listen and observe for experimentation and application of ideas; listen for comments and observations, responses to questions

Closure: Observe for results of experimentation; listen for responses to questions

VIBRATION

physics; n-k/primary

Purpose:
To determine sources of vibrations and begin associating vibrations with sounds

Thinking skills:
Science problem solving: Observation, data collection, data interpretation, communication of solutions
Creativity:
    Originality: Uniqueness of solutions
Communication: Listening, verbal, non-verbal, interpersonal

Vocabulary:
Vibration, vibrate, sounds, frequency, strum, strike

Prerequisites:
Prior contact with stringed and percussion instruments, such as guitars, ukuleles, autoharps, triangles, rhythm sticks

Materials:
Variety of musical instruments which children can manipulate, such as auto harp, guitar, triangle, cymbal, drum, and tambourine; rubber bands, plastic and/or wood 1 ft. and 1 yd. rulers, empty tissue boxes with center holes, combs, toilet paper, soda bottles partly filled with water; hard table tops; radio

Introduction:
Gather the children together and ask them to be completely quiet. Then ask that they place their finger tips on their larynx and feel what happens, first when they breathe, next when they hum and finally, when they talk. Spend several minutes discussing their observations and introducing the notion of "vibration." Tell them other things can be made to vibrate and they will be able to try on their own. Ask them to select several materials from the array and return to their tables (desks) to work.

Continuation:
Begin playing the radio softly in the background. Invite several children at a time to feel the speaker and discuss their observations. Also encourage remainder of the group to share materials, discuss their discoveries and help each other. After the radio has been shared with every child, begin circulating among the children and asking questions. Use vocabulary as appropriate.

Application:
How can you use the *(name of object)* to make a vibration?
What will happen if you *(name of action)* to the *(name of object)*?
Which object do you expect to make the most *(least)* vibration *(sound)*?
How will you get the bottle to vibrate?
What must you do to change the sound that comes from a vibrating object?

Analysis:
Why is there usually a sound when something vibrates?
How does a vibration work?
What generally happens when objects vibrate?
What differences in vibration occur in different objects?
What produces the sounds from a vibrating *(name of object)*?

Synthesis:
Compose a piece of "music," using your vibrating objects to play it.
How would you improve on the vibrations of a *(name of object)*?
If you could vibrate, how would you sound?

Evaluation:
Which object made the nicest *(ugliest)* sounds?
Which combination of vibrations is least *(most)* pleasant?

Closure:
Ask the children to select one source of vibrations other than a musical instrument and bring it to the large group gathering. Have everyone play their vibrations together and then spend a few minutes discussing the sounds produced.

You should ask everyone to help clean up after the discussion.

Learning check:
Introduction: Observe for careful participation and following instructions, examination of materials
Continuation: Observe and listen for attention to details, cooperation and mutual sharing; listen for comments and responses to questions
Closure: Observe and listen for participation in final solution and discussion

WATER PLAY WITH TUBES

physics; n-k/primary

Purpose:
To investigate the fluid properties of water and to control its action

Thinking skills:
Science problem solving: Problem finding, solution finding, solution testing, communication of solutions
Creativity:
  Originality: Unusual uses of materials
  Elaboration: Detailing of solutions
Communication: Non-verbal, interpersonal

Vocabulary:
Tube, funnel, sprinkler, insert (verb), flow, fluid, level

Prerequisites:
A variety of prior water play experiences

Materials:
Large tubs of water (approximately 10 gal. each, half filled for indoor use or 5 gal. buckets mostly filled for outdoor use); empty buckets, clear plastic tubing of many lengths and sizes chosen to fit snugly inside one another, sprinkler nozzles to fit tubing, corks to fit tubing, plastic measuring cups, funnels to fit into or onto tubing, squirt bottles of food coloring; utility knife for cutting tubing; mops, sponges, towels, smocks

Introduction:
Group children (3-4 for indoors, 2 for outdoors) to work together and put on smocks. Show the children the materials and tell them they may use the materials however they wish but should try very hard to keep the water in the containers. Let them get started exploring.

Continuation:
Observe how the children are using the materials. If needed, model how the pieces of tubing can be combined to create extended lengths, closed loops and the like. Ask questions and introduce vocabulary.

Application:
How can you use these materials?
What will happen if you *(name of action)*?
How do people use tubes and things like tubes?
How can you construct a multi-colored fountain?
What will the water do when you raise *(lower)* this *(indicate)* part of the tubing?

Analysis:
Why did the water do that *(indicate results of action)*?
What usually happens when you raise *(lower)* part of the tubing?
Why did *(didn't)* you get your fountain to work?
Why do the levels of the water in the loop always match?
What other conditions affect the level of water in the tubes?

Synthesis:
What do your tubing, funnels, and water remind you of?
How would you design a continuous fountain?

Evaluation:
What strategy made the best fountain?
Which part of the activity was most *(least)* interesting *(challenging)*?

Closure:
Give the children a two to five minute warning that they will have to finish their experiments and get their best fountains ready to show the rest of the class. At the end of the time, have each group demonstrate its fountain. Once demonstrations are complete, ask each group to place all materials in the water and mop up spills if necessary. Then they should all dry hands and return to seats (class).

Learning check:
Introduction: Observe for exploration of materials, cooperation within the groups
Continuation: Observe for uses of materials, continued cooperation; listen for responses to questions
Closure: Observe for successful solutions

WHEELS AND AXLES

physics; n-k/primary

Purpose:
To investigate the nature of wheels and the relationship between wheels and axles

Thinking skills:
Science problem solving: Solutions finding, solution testing, communication of solutions
Creativity:
   Originality: Uniqueness of solutions
Communication: Verbal, listening

Vocabulary:
Wheel, axle, disk, balance

Prerequisites:
Prior play with wheel toys with visible axles

Materials:
Wheels of various sizes from wheel toys, wagon wheels, bicycle wheels, lids from paper cups; wooden and/or plastic disks, squares, ovals, and irregular shapes, some with center holes, some with off-center holes; boards and blocks to construct ramps; straws, dowels, small pieces of pipe (to serve as axles); balls; large table tops or open floor space; experience chart materials

Introduction:
Ask the children to describe what a wheel looks like and how it works. Keep a record of their ideas, then ask what keeps wheels from tipping over. Give the children a chance to respond. Show them the materials and encourage them to experiment with making working wheels.

Continuation:
Observe to see how the children's experiments are progressing. Comment on their work, introducing vocabulary and asking questions as appropriate. If they are highly successful in constructing wheels with axles, introduce the balls into the materials.

Application:
How do people use wheels?
What will happen if you try to roll a *(name of object)*?
What will happen if you attach two *(name of objects)* together somehow?
What are you going to do with the straws *(dowels, pipes)*?
How can you test your wheel and axle design?

Analysis:
What happens when you put the axle in this *(indicate)* hole?
How are a ball and a wheel alike *(different)*?
Why did your wheel *(wheel and axle)* move like that?
What usually happens if your wheels are different sizes?
Why do *(don't)* you want a third wheel on your axle?

Synthesis:
How would you design a wheel and axle system for a smooth *(bumpy)* ride?
How can you adapt the balls so that they can be wheels?

Evaluation:
Which objects make the best *(worst)* wheels?
What was the easiest way to keep a wheel from tipping over?

Closure:
When it's time, ask the children to roll all of their wheels into one pile, then stack up the rest of their materials as you prefer. Once clean-up is complete, bring the children together to discuss their experiences. Compare comments to your chart, emphasizing characteristics of wheels and axles and how they work. If there is serious confusion, bring out wheel toys for the children to examine and discuss.

Learning check:
Introduction: Listen for participation in discussion; observe for experimentation
Continuation: Listen for responses to questions; observe for successful solutions
Closure: Observe for participation in clean-up; listen for contribution to discussion

BLENDING MEDIA

chemistry; primary

Purpose:
To blend media types to create new and interesting effects for painting

Thinking skills:
Science problem solving: Problem finding, observation, hypothesis formulation, hypothesis testing, data collection, data organization, data interpretation
Creativity:
   Fluency: Production of alternative trials
   Flexibility: Variety of alternative trials
   Originality: Unusual combinations
Communication: Non-verbal, listening

Vocabulary:
Gritty, speckled, dissolve, disappear, lighter, waxy, suspend, blend, darker, oily, suspension, emulsion, medium, media

Prerequisites:
Familiarity with all media used; experience blending secondary colors

Materials:
Tempera paint powder of various colors, water color cakes of various colors, pastel chalk, crayons, water soluble inks, food colorings, candle colorings or stubs of old candles, Easter egg dyes, vinegar (if needed for egg dyes); water, vegetable oil; 15-20 baby food or other small jars, empty tin cans, paper cups (DO NOT use styrofoam cups); old sauce pan, hot plate, potholder mitt, food grater, small kitchen knife, paper plates, wooden coffee stirrers; 15-20 paint brushes of various sizes, matte and glossy paper to paint on (several sheets per child); paper towels, rags, dish detergent, two buckets of warm water (for clean-up); well ventilated work space

   Provide sufficient amounts of materials for everyone to have an ample amount to experiment with; restrict colors to red, blue, yellow, black, white.

   Working with one small group (4-5) at a time, array materials in well ventilated work space so that they are accessible to all children. Place hot plate and old pot nearby for easy access. Be certain pot has water in it at all times and is restricted to low heat only.

Introduction:

Show the children the materials. Tell them they will have a chance to mix their own types of pigment in their own colors for painting. They may blend any materials they like and may heat them if they wish. Talk about some pictures they would like to paint and the colors and special effects they will need. Ask how they will make their colors and keep a chart of their initial ideas.

Review safety rules: always wear mitt when cooking; be certain there is water in the pot before placing a jar in it to heat; always heat jars in pot of water; do not put any materials in mouth.

Encourage each child to create interesting qualities and textures in their painting materials. Ask them to keep notes about what they do: combinations of media; colors used; heated or not heated.

Continuation:

As the children work on blending their colors and media, help them move beyond the usual combinations. Encourage them to mix media with very different properties, grate and slice hard media, warm media to melt waxes, etc. As the children work on blending textures and colors, they should paint test patches on a sheet of paper to determine if they are getting the results they want or if there are surprises in their media. These paint patches can be used as points of reference from which they might choose to modify their experiments to achieve new effects.

Refer to the questions below to help expand the children's thinking and experimentation.

Application:

How would you use the new medium you have created?

How would you use this color *(color combination)* in your painting?

Draw a sketch of the painting you plan to paint, showing the colors and textures you want to include.

How will you modify your planned painting as a result of your findings about the media?

How will you make the colors *(textures)* you want?

Analysis:

Why does this medium act as it does?

How did you get this color *(texture)*?

What will be the effect of combining these media?

What usually happens when you mix watery media and waxy media *(watery media and oily media)*?

What are the differences *(similarities)* among the media you have created?

How do you determine if you get the results you want?

Synthesis:

How can you organize these various media to show their differences and similarities?

How will you design your painting to use your new media?

What would be a good name for this medium *(color, texture)*?

Evaluation:

Which of the new media is *(are)* most interesting *(unusual)*?

If you had to work in only one medium, which would you choose?

How well do the new media meet your expectations and needs?

Closure:

Ask the children to complete their paintings and put them in an appropriate place to dry. Ask them to select those media which they would like to save and place lids on them. Then ask if any of the rest of the group is interested in saving media left to be discarded. When all media to be saved have been set aside, ask the children to try to wash their brushes in the buckets of warm water. Help them use the detergent to clean the waxy substances from the brushes. Once they have scrubbed the brushes, they need to set aside any they think may still have media in them (these will need to be heated in soapy water to clean). The children should then dispose of soiled newspapers, paper towels and the like. Once the dirty water is poured out, rags can be stored in the buckets.

It is important that the children be actively involved in clean-up since they will continue to make new discoveries about the media.

Once clean-up is complete, the children may wash their hands and take off their smocks. Ask the children to gather so that they can discuss the results of their work. Have them talk about how accurate their predictions were, whether they were able to make the colors and textures they wanted and how their paintings came out. Discuss their successes and failures as related to their predictions. Discuss any changes they made in their plans which resulted directly from their discoveries about the media. If any child made some particularly startling discovery, he might want to show his painting to the rest of the group. Plan some media combinations which they might like to try in the future.

Learning check:

Introduction: Observe for adherence to rules, record keeping, experimentation; listen for relevant descriptive comments

Continuation: Observe for continued adherence to rules, record keeping, new ideas, experimentation with materials; listen for comments about experimentation

Closure: Observe and listen for conclusions about observations

BOWLING

physics; primary

Purpose:
To explore relationships among objects used in bowling game, including arrangement of target objects and path of balls

Thinking skills:
Science problem solving: Data collection, data organization, data interpretation
Creativity :
   Originality: Uniqueness of solutions
   Elaboration: Refinement of solutions
Communication: Interpersonal, verbal, non-verbal

Vocabulary:
Target, aim

Prerequisites:
Prior experience with child's bowling game; knowledge of simple addition of sums to 10 will help

Materials:
Tin cans of different shapes and sizes, 2-liter plastic drink bottles, milk cartons, long and short hollow cylinders; square wooden blocks, long rectangular blocks; variety of soccer balls, tennis balls, softballs, and big plastic balls

   For each group of 2-3 children, set out a variety of 10 target objects (tin cans, plastic bottles, milk cartons, cylinders), a variety of balls, several of each type of block.

Introduction:
Tell the children that you have set out materials for them to use for bowling. Ask them to describe what they know about bowling and reinforce that the ball must stay on the floor during bowling. Explain that in this bowling game they are allowed to arrange the target objects however they wish and use whichever types of balls they wish.

   Ask them to choose or assign them to their "teams" and ask each team to get some paper and a pencil to keep records. Then instruct the teams to go to the sets of materials and begin playing as soon as they are ready.

Continuation:
Circulate among the children and observe how they are working with the materials and if they are attempting to keep "score" or other records of what they do. Focus their experiments by asking questions to guide their thinking.

Application:
How else can you make the ball hit the target?
What other ways can the targets be arranged?
What would happen if you put a bend in your bowling alley by using the long and square blocks?
How can you keep track of what you do each time you bowl?
If you ran a bowling alley, how would your arrange the games?

Analysis:
What are the differences *(similarities)* among your arrangements of targets?
How does each of the different balls work for bowling?
What is the effect of making a bend in the bowling alley?
What other games require balls to bounce off the sides of things and how are they like your bowling?
What will happen if the ball hits the side of your bowling alley?

Synthesis:
How would you design the ideal bowling alley?
What can you do with your information to share it with the other bowlers in our class?

Evaluation:
Which was the most *(least)* challenging arrangement of targets?
Which arrangement of targets, alley and balls would make people most want to go bowling?

Closure:
After 20-30 minutes, give the children a warning that in five minutes they will have to stop bowling. When the five minutes are up, ask the children to place all the materials in the boxes near where they are working. When everything is put away, ask them to come together with their record sheets to discuss the games they tried. Have each group share what they tried and how well it worked. If they kept score, ask what the scores were. An experience chart with pictures and scores would help other groups understand what was tried. Ask the group as a whole to draw some general conclusions about their experiments with target arrangements, types of balls, and bent alleys. Record their responses for future reference.

Learning check:
Introduction: Listen for comments about prior experiences, willingness to cooperate
Continuation: Observe for score keeping and other records, construction of solutions, team cooperation; listen for cooperation, sharing of ideas
Closure: Observe for record keeping; listen for understanding of experiences

CHAIN REACTIONS

physics; primary

Purpose:
To construct physical chain reactions using simple, common materials

Thinking skills:
Science problem solving: Problem finding, observation, hypothesis formulation, hypothesis testing, data interpretation
Creativity:
  Elaboration: Creating complex solutions
Communication: Verbal, non-verbal, interpersonal

Vocabulary:
Chain reaction, delay, spacing, flexible

Prerequisites:
Familiarity with building blocks

Materials:
Sets of dominoes, sets of identical building blocks or other identical rectangles of wood or plastic (they must be heavy enough to work in a chain reaction); rulers, paper, pencils; level table tops or smooth open floor space

Put about 10-12 identical blocks or the like in each set. Plan on enough sets for the children to work in pairs.

Introduction:
Set up a model chain reaction. Bring the children around the model and set it off. Tell the children you would like them to try to duplicate what you just did and hand each pair their materials. Let them discuss and experiment freely at first.

Continuation:
Observe their work with the materials and ask that they begin keeping records of what they are trying. Introduce vocabulary and ask questions as appropriate.

Application:
How do you plan to approach the problem?
What will happen if the spacing between blocks is changed?
What would happen if you mixed up all different sizes of blocks?
How can you keep records so you know what you've tried?
What would happen if your materials were very flexible?

Analysis:
What happened in the model chain reaction?
What happened when you tried to make a chain reaction?
What conditions are needed to get a chain reaction?
How would using more blocks affect a chain reaction?
Why does a chain reaction work?

Synthesis:
What does a chain reaction remind you of?
How would you design a bigger and better chain reaction?

Evaluation:
How well did your chain reactions work compared to the model?
What is the importance of understanding chain reactions?
What materials work best *(worst)* in a chain reaction?

Closure:
Ask the children to stop for a few moments and share their experiences. Have them tell about what worked and what did not when constructing a chain reaction. Then ask how they should combine all of their materials to create one very long chain reaction (or you may want to do several shorter ones). Let them actually set up the final chain and when they are ready, one child should set it off so everyone can see how well their ideas work. You may want to take a couple more minutes to discuss the giant chain before cleaning up.

Learning check:
Introduction: Observe for experimentation with materials; listen for relevant comments
Continuation: Observe for continued experimentation, evidence of solutions, record keeping; listen for cooperation, responses to questions
Closure: Listen for sharing of experiences; observe for participation in final solution

CLAY

physics; primary   designed by Susan Gayle Golden

Purpose:
To explore the plasticity of raw versus hardened clay

Thinking skills:
Science problem solving: Observation
Creativity:
   Fluency: Production of alternative trials
   Flexibility: Variety of alternative trials
   Originality: Uniqueness of alternative trials
   Elaboration: Detailing of alternative trials
Communication: Non-verbal, verbal, listening

Vocabulary:
Clay, pottery, mold, sculpt, sculpture, plastic (as used for clay), kiln

Prerequisites:
Extensive experience using plasticine (non-hardening) clays and playdough

Materials:
Large quantity of self-hardening (firing) clay; plastic knives, sticks, dowels (1 foot lengths), objects for making patterns; newspapers, smocks, paper towels or rags, buckets of water, empty buckets, lawn-sized garbage bags; glazes, paints, brushes (all optional); kiln (if needed); fired or dried pieces of pottery

Introduction:
With the whole group, review rules for working with clay: each child works on his or her own clay only; share knives, sticks, dowels and other objects. Ask each child to cover his or her desk with 5 or more thicknesses of newspaper and put on a smock. Each child should come up to see the materials available and get his or her piece (large) of clay. As you distribute the clay, explain that this is clay which can harden (be hardened) so that the children should be sure to keep it damp as they work. Tell them they may make anything they wish from the clay.

Once the clay is distributed, place the buckets of water in various locations around the classroom for the children to dampen their clay.

Continuation:
If there is extra clay, distribute it to the children who want to make larger objects. If you are using the kiln, take a few children at a time to the kiln, tell them about firing the clay and let them examine the kiln and the fired pottery. You may also want to talk a little about how pottery is glazed.

As the children work, circulate among them and ask questions about what they are doing and planning to do. Use appropriate vocabulary.

Application:

What can you make out of all this clay?

What are some things you've seen that are made out of hardened clay?

What will happen if you drop raw *(hardened)* clay?

How will water change the texture of your clay?

What will you have to do to make the clay into a *(name of object the child said he is trying to make)*?

Analysis:

How do the raw and hardened clay each feel?

What other materials feel *(look)* like raw *(hardened)* clay?

Why does the clay become hard when it sits out in the air *(when it is fired)*?

What happens when you press together two pieces of raw clay?

Why is it so easy to make the raw clay take any shape you want?

Synthesis:

What title will you give your finished clay sculpture?

How do you want to display your finished piece of work?

Evaluation:

What is the easiest way to make clay take the shape you want?

For making sculptures, how does this clay compare to our plasticine and playdough?

What do you like best *(least)* about working with this kind of clay?

Closure:

Give the children five minutes warning prior to clean-up. Ask them to be thinking about how they want to decorate their sculpted clay once it is hard. When it is time for clean-up, ask them to set aside sculpted clay in a designated place. Ask the children how a lump of clay might be helpful in cleaning up, and what will be the best ways to clean all the utensils and other materials. Let the children each experiment in his own workspace with clean-up strategies. Make empty buckets available for putting in excess clay and cleaned materials. Keep an eye out so large quantities of usable clay are not thrown out or placed in the buckets of water. Once all the materials are cleared away, the children should wad up the newspapers from their desktops and throw them into the garbage (a couple of lawn-sized garbage bags will make this easier). When the newspaper is off a child's desk, s/he should scrub hands and arms in a bucket of clean water, dry off, remove smock, and return to his or her seat. Or you may wish to let the children look at (but not touch) each other's work. It would be a good idea if at this point each child wrote down or drew a picture of how s/he will decorate the pottery so ideas are not forgotten. You may also want to discuss the activity as a whole group and plan what the children will look for as the clay hardens or when it is fired.

Learning check:

Introduction: Observe for following preparatory directions

Continuation: Listen for observations and other comments; observe for production of alternatives

Closure: Observe for successful solutions, following directions, attending to peers

COOKED SNACK

chemistry; primary   designed by Susan M. Brownell

Purpose:
To make up recipes for snacks with selected ingredients

Thinking skills:
Science problem solving: Problem finding, observation, hypothesis formulation, hypothesis testing, data collection, solution finding, solution testing, communication of solutions
Creativity:
  Originality: Uniqueness of solutions
  Elaboration: Detailing of solutions
Communication: Verbal, non-verbal, interpersonal

Vocabulary:
Ingredients, experiment, capacity, blend, utensils, texture, consistent, property(ies), thickness

Prerequisites:
Some simple experiences in cooking with recipes, such as making popsicles, baking cupcakes or cooking pudding

Materials:
Kitchen area with stove, oven, sink, counter space; pots of various sizes, potholders, bowls of various sizes, mixing spoons, measuring cups and spoons, baking bowl or pan, serving dishes for snack, beater/mixer, aprons, hot pads, spoons, paper, paper for recipe chart, pens/pencils; detergent, sponges or dish cloths, towels; fruit (canned, fresh or frozen), peaches, pineapple, fresh bananas, berries; containers of flour, sugar, salt, cornstarch, cornmeal, grain cereals (oatmeal, rice); eggs, milk, flavorings, cocoa, tapioca, water; cookbooks (optional)

Introduction:
Show the ingredients (listed in materials) to a group of 4-5 children and ask them what they might do with them. If no suggestions are made, ask "What can you do with these ingredients to make something you and your friends will like for snack?" Explain that they will be able to use and experiment with any of these ingredients in any combination they think will make a good snack (or whatever they have suggested).

Before the children begin cooking, review health and safety rules, including washing hands before handling food, cleaning up afterwards, and care using the stove and oven. If necessary, demonstrate safe use of the stove and oven.

Once rules are reviewed, ask the children to examine the ingredients closely and taste a little bit if they need to. Have them plan in advance what they would like to prepare and write down their ideas, including the amounts of each ingredient to be used, how ingredients will be blended, etc. The children should also plan how they will share the various jobs involved.

Continuation:
As the children work, you should be in the background asking questions when the group gets stuck or side-tracked. The children may need to be reminded to write down any observations they make and to keep records of changes they make in their original plans. You should help them keep focused on the idea that they are working toward an edible

result, asking them how it tastes and looks and would they enjoy eating it. Ask any of the following questions as they become appropriate to the children's actions and decisions.

Application:
How might you use *(name of ingredient)* in your snack?
What will happen when you *(add something, do something to the ingredients, etc.)*?
How do you think it will change as its temperature changes *(cools, heats up)*?
How much of each ingredient will you need?
How will you display your snack?

Analysis:
Why did your snack turn out as it has?
When did the mixture begin to change and how?
How will you decide when it is done cooking *(ready to eat)*?
What can you conclude about the properties of *(name of ingredient or name of mixture)*?
How should you organize your recipe so that someone else can prepare the same snack?

Synthesis:
What will your recipe for a snack consist of?
What name will you give your snack so everyone will remember it?
What other snack recipes might you develop using these ingredients?

Evaluation:
Is the snack what you expected it to be?
Do you want to share it with your friends or do you want to wait and make changes in the recipe?

Closure:
Have the children record their discoveries and decide what happened to the snack based on the preceding questions. If you desire, they could compare their recipe with one in a cookbook and decide how theirs could be improved, relating to the above questions and their answers. They may want to try again, especially if it burned or tasted unpleasant.

The group should clean up the utensils used while the snack is cooling (part of the preparation process). Each one might clean up a specific area, or type of utensil used. Once clean-up is completed, ask the group if they are ready to share their snack or if they made other plans for it. Proceed as appropriate to their response.

NOTE: This activity may lead into experimenting with the various ingredients and investigating which of the ingredients thickens liquids such as milk or water. Ask "How can we find out if each ingredient will get thick or make something thicker? What happens to eggs when they're cooked in a pan (out of the shell) (in the shell)"? The children may think of or be encouraged to try each ingredient separately - mixed with water or milk and heated. They could be guided to try different amounts of the liquid and the thickener and record the results. Ask "What happens if you have more flour than water? What happens if you have equal amounts? Or less flour than water"? (These ideas could also serve as plans for next time the group cooks.)

Learning check:
Introduction: Observe and listen for cooperation in planning, suggested possibilities or alternatives, written plans
Continuation: Observe and listen for cooperation, solutions to problem, written records, approach to a single best solution
Closure: Listen for group decisions; observe for single best solution including written record

GOLF

physics; primary   designed by Debbie Wilson & Angela L. Blough

Purpose:
To control path of hit ball through knowledge of angle of impact of "golf club" on ball

Thinking skills:
Science problem solving: Problem finding, hypothesis formulation, hypothesis testing, solution finding, solution testing, communication of solutions
Creativity:
  Fluency: Production of alternative trials and solutions
  Elaboration: Detailing of solutions
Communication: Verbal

Vocabulary:
Angle, aim, target, direction, left, right

Prerequisites:
Experience playing target ball using a backboard and sideboards

Materials:
Tennis, golf, ping pong, and small rubber balls, marbles; masking tape, long boards and blocks, large tin cans; varying (1-3 ft.) lengths of 1 in. x 1 in. boards, some with a 1 in. x 1 in. x 4 in. board attached to the bottom to make a "T"; large box, large playspace

Introduction:
Show the children the sticks and balls and ask them how they can use the sticks to move the balls across the floor. Before they make some suggestions, remind them of essential safety rules: balls must remain on the floor; sticks are to be used only with the balls. Once they have shared a few ideas, tell them they are free to experiment for a little while. Observe them carefully while they work, interacting as appropriate.

Continuation:
After about five minutes of free experimentation, set out the cans on their sides around the playspace, suggesting to the children that they try to get the balls into the cans using the sticks. Begin asking pertinent questions as they work. After they become fairly successful hitting balls into the cans at close range, introduce the remaining materials. Ask how they could use these things to make their golf more challenging. If you need to, remind them of how they played target ball or bowling games using boards to make lanes. Use questions to further their thinking.

Application:
How can you use these materials to make your game more interesting *(challenging)*?
What would happen if you used a shorter *(longer, T-shaped)* stick *(different kind of ball)*?
How can you use the stick to make the ball move?
What are some other ways to make a path for your balls?
How many times will you have to hit the ball to get it in the can?

Analysis:

Why did the ball go in that direction?

What makes the balls go faster *(slower, farther)*?

Why do the different kinds of balls act differently?

How did your path make the ball move?

What usually happens when you use a short *(long, T-shaped)* stick?

Synthesis:

What kind of new game can you make using these same materials?

What rules will you need for your new game?

How could you make golf a team game?

Evaluation:

Which materials made your golf game most challenging?

What was the most important change you made in your game?

Closure:

Take the large box and turn it on its side. Ask the children to hit all of their balls into the box, then pick up the rest of their materials. (You may want to use this box as storage for all of the equipment.)

Once clean-up is complete, bring the children together to share how they played. Also, ask how they will make new games the next time they golf. You might want to write down (draw pictures of) their ideas.

Alternative:

When clean-up is complete, the children could return to their seats and write down or make pictures of the games they invented. Older children could also try writing rules for their games.

Learning check:

Introduction: Observe and listen for exploration of materials; listen for comments on observations, explanations

Continuation: Observe and listen for experimentation with materials, demonstration of solutions

Closure: Listen for suggestions

LEVERS AND FULCRUMS

physics; primary

Purpose:
To construct and utilize levers to move large objects

Thinking skills:
Science problem solving: Observation, problem finding, hypothesis formulation, hypothesis testing, solution finding, solution testing, communication of solutions
Creativity:
   Flexibility: Variety of alternative trials
   Elaboration: Detailing of solutions
Communication: Verbal, interpersonal

Vocabulary:
Lever, fulcrum, weight, position, leverage, force

Prerequisites:
Prior experience with some of the following: seesaw, pry bar or crow bar in woodworking, pry-type bottle opener, long blocks in building, long-handled shovels

Materials:
Several very heavy, awkward objects which can be moved by levers (logs and large rocks are best but make sure they cannot be rolled by 2-3 children working together); boards of varying lengths (1 ft. - 10 ft.) and varying cross-sections (slats, 1 in. x 1 in., 1 in. x 2 in., 2 in. x 2 in., 2 in. x 4 in., dowels) (be certain at least one board is adequate to move object using lever construction); fulcrums - various diameters of metal and/or plastic pipe, short pieces of 1 in. x 1 in., 2 in. x 2 in., 4 in. x 4 in., 6 in. x 6 in., miscellaneous smaller rocks and/or logs; hand shovels; outside space to work (preferably where digging is possible - a sand pit is ideal)
   Plan your materials for 1-5 children working in each group. Place logs or rocks outdoors to create a problem situation. For example, the logs or rocks need to be moved out of the way because they are obstructing a play space. Placing one or two in depressions in the ground will make the task more complex. Make other materials readily available in the immediate vicinity.

Introduction:
Take the children to the area where the ''problem'' exists and encourage them to examine the situation and materials. Bring them together into a group and discuss plans of action, being sure NOT to tell them about lever construction.
   Some questions you might ask are:
      What are the circumstances of our problem?
      What do you already know about solving this problem?
      What are some possible uses of the materials we have available?
      What are some of the relationships among the various materials?
   Have the children organize themselves for the tasks. Let them group themselves or work alone as they choose. Ask them to select an object they believe they can move and then to plan their strategy before attacking the problem physically. Once they have discussed their ideas and are ready to try them, the children should proceed.

then to plan their strategy before attacking the problem physically. Once they have discussed their ideas and are ready to try them, the children should proceed.

Continuation:

Observe the progress of the various groups and individuals and ask questions from below as appropriate. As you discuss things, introduce correct vocabulary. Encourage the children to observe when a promising solution is in progress.

If levers are readily constructed and used, add to the problem by requiring a log to be stood on end or a log or rock to be rolled in a straight path to a selected location. Both tasks will necessitate coordinating 2 or more levers, as will extracting an object from a depression in the ground.

Application:

What can you use this piece *(lever or fulcrum material)* for?
How much force will you need to move the *(object)*?
How might you use two levers working together?
What will happen if you try *(specify action)*?
What are some other ways you can use levers?

Analysis:

How do the lever and fulcrum work together?
What happens when you change the lever *(fulcrum)*?
What are some of the factors which make this problem difficult *(easy)* for you to solve?
How does a lever work?
Why do you think your idea did *(did not)* work?

Synthesis:

Suppose we had a log/rock three times this size; what would you have to do to move it?
How would you do things differently if you were to try to solve this problem again?
How do the different combinations of levers and fulcrums make work easier or harder?

Evaluation:

Which was the most *(least)* effective technique you used to move the rock/log?
Of the strategies you have not tried, which do you suppose would be the best and why?
What do you think is the most important use for levers?

Closure:

When all the groups have developed levers and have moved their objects successfully, request the children to make a neat pile of the levers and a pile of the fulcrums. Then bring the children together to discuss their experiences. Ask each group to describe what it did and how. Use a written record or drawing of their experiences so comparisons can be made among solutions. Discuss similarities and differences among solutions and use questions from above to guide the discussion.

Learning check:

Introduction: Listen for group collaboration, discussions of observations and ideas
Continuation: Observe for experimentation with materials; listen for sharing experiences, explanations of discoveries, cooperation
Closure: Listen for communication of solutions

MIRRORS AND LENSES

physics; primary

Purpose:
To control the focus of images created by lenses and mirrors

Thinking skills:
Science problem solving: Problem finding, observation, data collection, data organization, data interpretation
Creativity:
   Originality: Creation of unusual visual images;
   Elaboration: Detailing of visual images
Communication: Verbal, interpersonal

Vocabulary:
Reflect, reflection, focus, concave, convex, image, inverted

Prerequisites:
Prior use of magnifying glasses and mirrors

Materials:
Assortment of flat, concave, and convex lenses; assortment of flat, concave and convex mirrors; large sheets of white paper, pencils, prepared charts for various lenses, mirrors and their images; well lighted workspace

CHART: Make a matrix with types of lenses and mirrors down left side and columns in which sketches of images at various distances (or upright and inverted) can be placed. You may want to draw pictures of as well as label lens and mirror types.

Introduction:
Bring the children together and ask what they see when they look into a mirror. Ask how the image acts and what it looks like. Also ask what happens when another person walks in front of the mirror. Inquire if any of the children remember using magnifying glasses and how they worked. Discuss their prior experiences and ideas as long as necessary.

   When the discussion is through, have the children team up in pairs and select some of the materials to investigate. Be sure that each team gets a chart to complete.

Continuation:
Observe their use of the materials and ask questions to stimulate further thought, introducing vocabulary as appropriate.

Application:
How do you plan to test your materials?
What will happen as you move the lens nearer to *(farther from)* the object?
What will you have to do to change the nature of the image?
How do people use different kinds of mirrors *(lenses)*?
What will you do with the records you are keeping?

Analysis:

How are the mirrors and lenses alike *(different)*?
Why do images go out of focus?
What happens to the image when you use a *(name of object)*?
What usually happens when you change the position of the lens *(mirror)*?
Why are some of your images up-side-down?

Synthesis:

How would you create an unusual image using lenses and mirrors?
What new use is there for inverted *(up-side-down)* images?

Evaluation:

Which images were the clearest?
What are the most important uses of lenses and mirrors?

Closure:

Give the children a five minute warning, then ask them to bring their charts to a group discussion. Ask them to share what they saw and learned, based on their charted data. They may then want to display their charts in a central location. When the discussion is over, the children should all help clean up.

Learning check:

Introduction: Observe and listen for participation, exploration of materials, beginnings of record keeping
Continuation: Observe for successful team work, record keeping, experimentation with materials
Closure: Observe and listen for participation in final discussion, clean-up

PENDULUMS

physics; primary

Purpose:
To construct pendulums and control the periodic rate

Thinking skills:
Science problem solving: Observation, hypothesis formulation, hypothesis testing, data collection, data organization, data interpretation
Creativity:
  Flexibility: Variety of alternative trials
Communication: Verbal, listening, interpersonal

Vocabulary:
Pendulum, period, bob, length

Prerequisites:
None

Materials:
String, bobs of various sizes, shapes and weights; frames or horizontal bars for hanging pendulums; scissors, paper and pencil, stop watches, balance scale, chart materials or blackboard

Introduction:
Group the children in teams of 2-3. Display a model pendulum and tell them they will have an opportunity to make pendulums like yours. Let them select the materials they think they will need and ask them to keep records of what they do. Observe as they begin to work.

Continuation:
Answer questions and offer assistance as needed with knots and the like. Begin asking questions and using vocabulary as appropriate.

Application:
How will you use the different materials?
What will happen if you shorten *(lengthen)* the string *(change the bob)*?
How will you keep track of what you have tried?
How do people use pendulums?
What do you expect will happen when you push *(drop)* the bob?

Analysis:
Why did your pendulum swing like that?
What happened when you shortened *(lengthened)* the string *(changed the bob)*?
What will you have to do to make your pendulum swing at the same rate as mine?
What usually happens when you push *(drop)* the bob?
How are our pendulums alike *(different)*?

Synthesis:

How would you design a machine which runs on a pendulum?
What kind of game can you invent using pendulums?

Evaluation:

What was the best way to make our pendulums swing at the same rate?
Which object makes the best bob?

Closure:

Ask each group to time the period of its pendulum(s), measure the length of string(s) and weigh bob(s). Using chart materials (or black board) make a display of this data, starting with shortest period and working up to longest period. Ask questions to encourage the children to associate the periodic rate with the length of the string, but do not try to force comprehension.

Learning check:

Introduction: Observe for attempts to reproduce model; observe and listen for cooperation, collaboration and record keeping

Continuation: Observe and listen for experimentation with materials, cooperation, record keeping, responses to questions

Closure: Observe and listen for following instructions, record keeping, participation in discussion

PENDULUMS WITH MULTIPLE TARGETS

physics; primary   designed by Martha A. Sweeny

Purpose:
To set up conditions for chain reactions using pendulums and targets

Thinking skills:
Science problem solving: Observation, problem finding, solution finding, solution testing, communication of solution
Creativity:
   Fluency: Production of alternative trials
   Flexibility: Variety of alternative trials
   Elaboration: Combination of multiple solutions
Communication: Verbal, listening, interpersonal

Vocabulary:
Chain reaction, pendulum, weight, speed, force

Prerequisites:
Prior experience with chain reactions and with pendulums and targets

Materials:
Pendulums (with interchangeable bobs heavy enough to knock over heaviest of target items); several identical of each item: wooden blocks, plastic bottles, milk cartons, tin cans; tennis balls, golf balls, pool or croquet balls, styrofoam balls.

Set up pendulums so they are tall enough and heavy enough to knock down target materials. They might be used outdoors tied to climbing equipment or indoors on rods supported between two desks or the like.

Introduction:
Show the pendulums and target materials to the children and ask them what they can think of to do with the materials. Make a list of their suggestions and then let them work alone or in groups to try out their ideas. As they experiment with targets, ask them what they should do to get the target objects to knock each other down. If a child or group has not yet begun to use targets, encourage them to observe children who are using the targets.

Continuation:
As the children work toward using multiple targets, introduce the idea of a chain reaction. Ask questions to further stimulate their thinking.

Application:
How could you arrange the objects so you could knock them down in a straight line?
What will happen if you use objects that are all the same *(all different)*?
What will happen if we move the objects farther apart *(closer together)*?
What will happen if you *(change the pendulums in some way)*?
How can you use several pendulums to make comparisons?

Analysis:

What happens when the pendulum hits the *(name of object)*?

Which objects will have a reaction similar to *(name ones being tested)*?

How are the reactions of the *(name of object)* and the *(name of other object)* similar *(different)*?

How can you make the targets move without knocking them down?

What actually happens in a chain reaction?

Synthesis:

How would you design a chain reaction machine?

What does your chain reaction remind you of?

Evaluation:

Which of the objects made the best *(worst)* chain reaction?

What is the most important thing people can use chain reactions for?

Closure:

Let the children work about 20-30 minutes on their experiments. Tell them that they will be cleaning up soon, but first you would like them to work together to make one very long chain reaction. As they work, ask them about the things they need to consider while constructing the chain. Be sure to talk to each child to make sure they each understand what is happening. Also make sure each child has a chance to help construct the chain. (If your group is very large, you might make two or even three long chains instead of only one.)

Once the children have succeeded in getting a long chain reaction to work, ask them to put the target materials where they belong and to take down the pendulums. If you desire, have a large group discussion about their experiences.

Learning check:

Introduction: Observe and listen for observations, idea sharing, experimentation with materials

Continuation: Observe for continued experimentation with materials, demonstration of solutions

Closure: Observe and listen for collaboration, solution, sharing, cooperation

PULLEYS

physics; primary

Purpose:
To develop systems of pulleys using various sizes and combinations to lift heavy weights

Thinking skills:
Science problem solving: Hypothesis formulation, hypothesis testing, data collection, data organization, data interpretation, communication of solutions
Creativity:
  Elaboration: Detailing of alternative trials
Communication: Verbal, non-verbal

Vocabulary:
Pulley, block and tackle, gravity, weight

Prerequisites:
Prior experiences moving weights in a variety of ways

Materials:
Various sizes of pulleys including several double pulleys; at least one complete block and tackle rig; long pieces of string of several sizes and types; long pieces of fabric (not plastic) rope with test strength of 100 pounds or more; various weights up to 50 pounds, such as buckets of sand saturated with water, concrete forms with screw eyes in them, window weights, buckets of rocks, empty buckets; climbing equipment, low tree limb, closet rods, or other strong, secure horizontal bars 4-5 or more feet high; pocket or utility knife (to cut rope and string); chair or step stool to climb on; board (optional); paper, pencils, crayons or markers

Hang at least one or two pulleys from a horizontal bar. Place weights underneath and other materials nearby. If possible, have a shelf built from the board approximately at eye level for a child.

Introduction:
Bring several different sized pulleys to the children in a small group and ask them what they know about pulleys and how they work. Discuss very generally and ask if they have ever seen pulleys in use. Tell them they will have a chance to experiment with different types of pulleys for lifting heavy weights. Ask them to make drawings of what they try as they work and to write what weights they could lift with their systems.

Continuation:
Observe progress with the pulleys and ask questions as appropriate while reinforcing vocabulary.

Application:
How could you use the double pulley *(block and tackle)*?
What are some ways pulleys could help us at home *(at school)*?
Draw and label a picture to show how your pulley system worked.
Please show another group how to thread the block and tackle.
When might you want a small weight to be able to lift a large one?

Analysis:
How do you think a block and tackle should be threaded?
Why do you think your system did *(did not)* work?
What else might work like a block and tackle?
What are the differences *(similarities)* among the pulley systems you have tried?
What kind of system would you need to lift a person *(a car, a house)*?

Synthesis:
How could you combine pulley systems to lift two weights at one time?
How would you build a system which uses a small weight to lift a much larger one?
How could you improve on the pulley system you are now operating?

Evaluation:
Which pulley system was easiest *(hardest)* to operate?
What would be the best way to decide how much weight a pulley system can lift?
How much easier is it to lift weights by using one pulley, two pulleys, more than two pulleys?

To advance understanding, after several experiences with pulleys encourage the children to experiment further. Have them try to develop a system in which two weights suspended at opposite ends of a pulley system hold up a much heavier weight between them.

Closure:
Gather children together around pulley apparatus and let each group explain and demonstrate its most effective pulley system. If any group made a surprising discovery, encourage them to share it. Children should try to explain why each system worked as it did. Return to discussion of pulleys in everyday life. Focus on how they might be used in machinery or when constructing tall buildings or bridges.

Ask the children to help clean up materials at the end of the activity.

Learning check:
Introduction: Observe and listen for attempts at explanations and demonstrations; observe for readiness to keep records
Continuation: Observe for experimentation with materials, record keeping
Closure: Observe and listen for successful solutions

SATURATES

chemistry; primary

Purpose:
To investigate how a material can have the properties of both a liquid and a solid

Thinking skills:
Science problem solving: Problem finding, observation, hypothesis formulation, hypothesis testing
Creativity:
  Fluency: Production of alternative trials
Communication: Verbal, interpersonal

Vocabulary:
Solid, liquid, semi-solid, appropriate words for tactile and visual descriptions, such as gooey, sticky, shiny, dull

Prerequisites:
Knowledge of the properties of solids and liquids

Materials:
Saturates and solutions of cornstarch and water (mix cornstarch and water in three different portions, each of which behaves as a liquid when poured and solid when squeezed, each tinted with a different food color); clear plastic cups, paper or styrofoam plates, plastic spoons; newspapers, paper towels; prepared charts of characteristics of solids and liquids, pencils

Introduction:
Have the children cover workspaces with newspapers and get a supply of paper towels, paper plates, plastic cups and spoons, and charts. Give each group of 3 children containers of each of the mixtures and ask them whether the substances are liquids or solids and what they think the substances are made of. (You may want to review properties of liquids and solids first, if you think the children may be unsure.) Encourage them to find as many ways to test their ideas as possible and to keep records on their charts.

Continuation:
Observe how the children are working, providing more of a given substance as needed. Listen for their observations and ask questions as appropriate.

Application:
What will happen if you mix the *(identify by color)* and the *(identify by color)*?
What will happen when you squeeze *(pour)* the *(identify by color)*?
How could you use these materials?
How are you going to test the *(identify by color)*?
What will happen if this stays in your hand *(on the plate)* for a long time?

Analysis:
Why do these substances behave as they do?
If the substances were all white, how would you tell them apart?
What do you think the substances consist of?
How would you classify each of the substances?
How are each of the substances like *(different from)* liquids *(solids)*?

Synthesis:
What do these substances remind you of?
How would you design a new product which uses any one of these substances?

Evaluation:
Which of the three substances has the most *(least)* desirable characteristics?
Which substance is most *(least)* like a liquid *(solid)*?
Which characteristic of the substance*(s)* is most interesting *(puzzling)*?

Closure:
After the children have been experimenting and marking their charts for enough time to have made decisions about the substances, ask that they please clean up. Everything should go in the garbage except charts and pencils. Then they should mop table surfaces and wash their hands.

Once clean-up is complete, gather to share solutions. You could compile ideas onto a wall-sized chart so everyone's findings can be seen together. Be sure to have the children discuss why they felt the substances were solid, liquid, both, or neither (or magic!).

Learning check:
Introduction: Observe and listen for exploration of materials, related comments, and group collaboration
Continuation: Observe for experimentation with materials. record keeping; listen for descriptive vocabulary
Closure: Observe and listen for solutions and explanations

SHADOWS

physics; primary   designed by Susan I. Petracca

Purpose:
To investigate how shadows are made and how their shapes are affected by the nature
of objects and light sources

Thinking skills:
Science problem solving: Hypothesis testing, data collection, data organization, data
    interpretation
Creativity:
   Flexibility: Variety of alternative trials
   Originality: Uniqueness of alternative trials
   Elaboration: Detailing of solutions
Communication: Verbal, interpersonal, listening

Vocabulary:
Opaque, transparent, angle, surface, texture

Prerequisites:
General awareness of shadows in the natural environment

Materials:
Bright flashlights; clear drinking glasses or clear plastic cups, cans of various sizes, tennis
balls, other small balls, several styles of hats; large sheets of paper, pencils, scissors;
various surfaces on which to make shadows (optional); large wall space in an area that
can be partially darkened

Introduction:
If it is sunny, take children for a short walk to observe the shadows created by the sun.
Ask the children to identify the objects creating the shadows. Ask them what is needed
to create a shadow. (If it is not possible to take a walk outdoors, tour your classroom
and school building for shadows.)

    Upon returning to the classroom, ask the children, working in groups of 3-4, to tape
the paper to the walls. Then ask them what they will need to do to make shadows of
parts of themselves on this paper. Accept their suggestions, and let them try to make
shadows, having a companion trace any shadows formed.

Continuation:
Once they have started making and tracing around the shadows, begin asking questions
and introducing other objects as sources of shadows. Introduce vocabulary as appropriate.

Application:
What will happen to your shadow as you move the light around *(nearer, farther away)*?
How will you make your shadow larger *(smaller)*?
What are some ways people might use shadows?
How can you change the shape of a shadow?
What will happen to the shadows if you turn out your flashlight?

Analysis:
What is the relationship between the light source and the size of a shadow?
What information about an object can you get from its shadow?
How does the movement of an object affect its shadow?
What kinds of shadows do you get from objects you can see through?
How would the shadow change if you filled the cup *(glass)* with *(water, milk, soda, chocolate milk)*?

Synthesis:
How could you combine several shadows to create an imaginary object?
How might you organize your shadow cut-outs to show several characteristics of the shadow's source?

Evaluation:
Which shadow is your favorite?
What was the best way to make a shadow of yourself?
Which was the most difficult shadow to make?

Closure:
After 20-30 minutes making shadows, ask the children to finish tracing and cutting the shadow they are working on. Ask them to share briefly what they learned. Also, ask them to relate what they learned about the length of a tree's (their own) shadows outside and the position of the sun. When the discussion is completed, have the children display their shadows on a bulletin board; they should select their most interesting and unusual shadows. They will then be ready to clean up their work spaces.

NOTE: Further investigation of shadows should involve various textures and contours of surfaces on which the shadows fall. These surfaces can be added during continuation above or used in a second activity involving shadows. It will be important to ask about the effects these surfaces have on the appearance of the shadows.

Learning check:
Introduction: Listen for cooperation and planning
Continuation: Observe for experimentation; listen and observe for group participation, tangible evidence of experiments, responses to questions, comments
Closure: Listen for narration of experiences and ideas; observe for contribution to display

SIPHONS

physics; primary

Purpose:
To construct and utilize siphons to move water between containers

Thinking skills:
Science problem solving: Problem finding, hypothesis formulation, hypothesis testing, data collection, data organization
Creativity:
   Originality: Uniqueness of alternative trials
   Elaboration: Refinement of alternative trials
Communication: Verbal, non-verbal, interpersonal

Vocabulary:
Siphon, suction, transfer, pump, gravity

Prerequisites:
Prior experiences with tubes and funnels in water play

Materials:
Several gallons of water in wide-mouthed, stable containers, such as 5 gallon buckets; narrow-mouthed containers, such as gallon jugs (several per group); plastic tubing of varying lengths to 4 feet and diameters to about ¾ inch interior, empty syringes, funnels, sand (optional); fish tank or clean gas can siphons, turkey basters; measuring cups, food coloring (to tint water to make it more visible), stop watches (optional), paper and pencils; paper towels, sponges, mops; sturdy tables or counter tops

   Set out materials in a roomy work area where water spills will be acceptable. Plan for groups of 2-3 children.

Introduction:
Gather the children together. Tell them there is a problem which needs to be solved. You have some large buckets filled with water and you need to put the water in the smaller containers. Tell them you would like them to find a fast way to transfer the water as *neatly* as possible. Permit them to color the water so they can see it better. They should be grouped, or group themselves, in work teams of 2 or 3 before beginning to solve the problem.

Continuation:
Observe the children and ask them questions from below. If they are having problems, it will be necessary to model a siphon. First, simply put the full container up on the table and the empty containers on the floor. Ask how they might use plastic tubing to transfer the water. If they continue to have difficulties, connect a pump to the tubing and involve the children in experimenting with operating the pump. Encourage them to help each other as much as possible.

Application:

How can you use the tubing *(turkey basters, funnels, pumps, measuring cups)* to help solve the problem?

What could these *(display but do not name the pumps)* be used for?

How will you keep records of the things you try?

What will happen if you *(name of action)*?

How long will it take to move the water using *(name of material/strategy)*?

Analysis:

Why is it difficult to pour water directly into the containers?

What are some problems with trying to use a *(name of object)* to transfer the water?

How does a siphon work?

Why do you have to put your finger over the end of the tubing *(pump)* to get the siphon started?

What are the advantages and disadvantages of using a siphon?

Synthesis:

How would you design an easy-to-use siphon?

Is there any way you could incorporate a turkey baster *(measuring cup, funnel)* into your siphon system?

Evaluation:

Did your siphon work as well as you would like?

Which was the neatest *(fastest, slowest, messiest)* of the strategies you tried for filling the narrow-mouthed containers?

What are the most important uses of siphons?

If the children construct siphons easily, introduce sand into the buckets and have them find ways in which the siphon can help them draw off the water without the sand. Ask them about the processes they try and their satisfaction with what they have done.

Conclusion:

When the children have succeeded in constructing operating siphons, encourage them to use the siphons to help clean up. They could siphon all the water into buckets or a sink. They should mop their workspace and clean up any spills on table or counter tops.

Once clean-up is complete, have the children gather to discuss their activities. Using questions from above, have them explain what they have done and why, displaying any illustrations they made while they worked. Ask the children how siphons can be used to solve everyday problems. If they have tried using siphons with the sandy water, encourage them to think about ways to use siphons to clean things.

Learning check:

Introduction: Observe and listen for cooperation, exploration of materials

Continuation: Observe and listen for experimentation with materials, cooperation, record keeping

Closure: Observe and listen for successful participation in clean-up, sharing of experiments

SOLUTIONS AND SUSPENSIONS

chemistry; primary

Purpose:
To create examples of solutes, solvents, and suspensions with common materials

Thinking skills:
Science problem solving: Data collection, data organization, data interpretation, solution finding, solution testing, communication of solutions
Creativity:
  Elaboration: Detailing of records and solutions
Communication: Verbal, interpersonal, listening

Vocabulary:
Solute, solvent, saturate, dissolve, suspension, mixture, substance

Prerequisites:
Understanding that substances behave in a variety of ways when mixed with other substances

Materials:
Liquids: water, vegetable oil, baby oil, alcohol, mineral oil, liquid detergent, vinegar, corn syrup, fruit juice, all in labeled containers (use at least three);
Solids: Polydent (Alka Seltzer tablets also work), aspirin, fizzing powder (like bromoseltzer), flour, sand, grits or corn meal, salt, sugar, powdered cleaners, powdered detergents, concrete or plaster of paris, baking soda, baking powder, corn starch, pepper, powdered chocolate, instant tea or fruit drink, jello powder, pudding mix, all in labeled containers (use at least three);
Clear plastic cups, measuring cups and spoons (optional), plastic spoons or coffee stirrers; newspapers, paper towels

Make sure that each group of 3-4 children has an ample supply of cups and stirrers to make all possible combinations of solids and liquids.

Introduction:
After establishing children's working groups, hold a discussion with them to introduce the activity. Ask the children if they have ever mixed instant drinks at home. Have them describe what they remember happening with the liquid and powder when it was mixed. Also talk with them about any cooking experiences they have had in which liquids and solids were mixed. Encourage them to share any observations they may have made during the cooking experience or any other occasions when they mixed solids and liquids (such as water in the sandbox).

Once they have shared their ideas, tell them that they will have a chance to experiment with mixing solids with liquids. Have each group collect its supplies and move to its work area. Instruct the children to cover work areas with newspaper. Then they may begin experimenting with mixing solids in the various liquids. Have them keep records of their experiments, using either a log book or a prepared chart of some type.

Continuation:

Circulate among the students. Discuss what they are doing, introducing and defining the vocabulary as appropriate. Ask the questions below to encourage further experimentation and more complex thinking skills.

Application:

How will you use this *(indicate)* material in your experiments?
What are some uses for your solution *(suspension)*?
What can you do with saturated solutions?
What will happen if you add this *(name material)* to your mixture?
What will happen if you add a whole lot of *(name of substance)*?

Analysis:

Which liquids have similar characteristics when mixed with other substances?
What are the characteristics of solutes *(solvents, suspensions)*?
Why might we need solutes, solvents, and suspensions?
Why did this solution saturate?
What do your records show about *(name of substance)*?

Synthesis:

Develop a plan for testing a new liquid *(solid)* to determine how it behaves when mixed with other substances.
How can you display your findings for the other groups to see?
Design a new use for one of your solutions *(suspensions)*.

Evaluation:

Which substance was the best *(worst)* solute *(solvent)*?
Which solution *(suspension)* was easiest *(hardest)* to create?
What is your favorite solution?

Conclusion:

As the children reach meaningful conclusions about the materials they are using, have them double check their records to make sure they tested all of their substances. Give them a several-minute warning that they will need to finish their experiments and decide what is most important in what they learned. Ask them to come together for a discussion and have them bring to the group discussion their best examples of a solution, suspension and saturate (if they have one).

Ask each group to share its most important finding and to identify its best examples. List examples on a chart so they can be compared. Groups should explain why they made their choices. Be sure to reinforce vocabulary during this discussion.

NOTE: During continuation, you might want to introduce one or two new substances as groups successfully create solutions, saturates and suspensions. Select substances which are similar to what the children are using so they can make predictions about them before testing them. If further extension is desired, introduce substances which will behave differently than those already used. Be certain to encourage complete testing of new substances.

Learning check:

Introduction: Observe and listen for participation, following instructions, cooperation with peers, experimentation and record keeping

Continuation: Observe and listen for continuing cooperation, responses to questions, record keeping

Closure: Observe and listen for sharing of experiences and solutions

SUDSING PRODUCTS

physics; primary

Purpose:
To establish criteria and test the quality of various sudsing products

Thinking skills:
Science problem solving: Problem finding, hypothesis formulation, hypothesis testing, data collection, data organization, data interpretation, communication of solutions
Creativity:
    Fluency: Production of criteria and alternative trials
    Flexibility: Variety of criteria and alternative trials
    Originality: Uniqueness of criteria
    Elaboration: Detailing of solution presentation
Communication: Verbal, interpersonal, listening

Vocabulary:
Criterion, criteria

Prerequisites:
Awareness of advertising claims about the qualities of various sudsing products; experience blowing bubbles in solutions

Materials:
Several liquid and powdered sudsing products; magazine/newspaper ads for the same and similar products arranged in a display; clear plastic cups, warm water, straws, stirrers, stop watches (one for each group), measuring cups and spoons; newspapers, paper towels

Introduction:
Show the entire group of children the ads and solicit their opinions about the ads and the products they represent. Identify the products available and ask the children about truth in advertising and how they might determine which ads are true and which are not. Also, ask them to suggest possible criteria (explain meaning if needed) for determining which product is best. Keep a chart of their suggestions of criteria for testing the products.

Once this discussion is complete, ask the children to form working groups of 3-4 (or assign them to groups). Have each group plan which products it will test and for what purpose. When their plans are complete, ask them to select the materials they will need, cover their work area with newspapers, and begin their experiments. Make sure each group keeps written records of plans, procedures and results and makes charts or displays of its data.

Continuation:
Once each group has gotten started with its experiments, begin circulating to see how they are doing. Ask questions when appropriate to clarify their ideas and activities and to encourage further testing. Use and reinforce the vocabulary.

Application:

What strategies are you planning to use to test your product(s)?
How can you use the information you are gathering?
How do you plan to keep your records?
What will you do if two products tie for quality?
What should you do if a product fails to meet advertised claims?

Analysis:

How might you interpret your data?
Why would a manufacturer advertise his product in this way *(refer to specific ad)*?
What criteria are needed to test your products?
Compare the usefulness of your various products.
What have you generally found about product advertisements?

Synthesis:

How would you improve on the products you have tested?
How would you advertise your best product?
What ways could you display your findings?

Evaluation:

How well does product *(specify)* quality compare to advertised claims?
Which criterion is most important or valuable for judging your product?
Which advertisement proved to be most accurate about product qualities?

Closure:

When the children have completed their testing of the products, ask them to clean up their work areas. They should then put the finishing touches on any charts or other displays of data. They should also draw up their ads for their best products. Then have them create two exhibits, one of data charts and one of ads.

Learning check:

Introduction: Observe and listen for participation in discussions, planning, cooperation
Continuation: Observe and listen for experimentation, record keeping, continued cooperation, responses to questions
Closure: Observe for successful participation in final displays

COMMERCIAL TOYS AND GAMES

Airhead
Balls
Block Builders Magnetic Blocks
Bocce Lawn Bowling
Bubblos
Cats Eye
Don't Break the Ice
Exit
Foam Hockey
Frisbees
Golf
Hungry Hungry Hippos
Juggling
Labyrinth
Leverage
Marble Mazes
Nerf Ping Pong
Paddle Ball (hard rubber and foam)
Pig Pong
Ping Pong
Playdoh
Pool
Poppin Top
Rebound
Shuttleball
Skee Ball
Slinky
Stuff It
Table Soccer
Tiddly Winks
Tip-It
Topple
Velcro Darts
Wee Waffle Blocks
Windup Propeller Planes

Ants in the Pants
Belly Bumper
Blockhead
Boomerangs
Building Blocks of any type
Construx
Don't Spill the Beans
Foam Croquet
Foam Ring Toss
Glider Planes
Horse Shoes
Ice Hockey
Kerplunk
Lawn Disks
Magnestix
Mousetrap
Nerf Pool
Pick Up Sticks
Pinball Games
Plastic Bowling Sets
Playdoh Fun Factory
Popoids
Pull Toys
Ring Toss
Sit 'N Spin
Skittle Bowl
Snap Bowling
Superfection
Ta-Ka-Radi
Tinkertoys
Tipsy Tower
Toss Up
Water Fueled Rocket
Wham O Rang
Zimm Zamm

LOGIC GAMES

Blockage
Down the Tubes
Kensington
Megiddo
Picture Triominoes
Superfection
Triominoes

Connect Four
Jigsaw Puzzles
Leverage
Pente
Score Four
Think & Jump